T0305343

Waste Management and the Green Economy

Wealth Inequality and the Green Economy

Waste Management and the Green Economy

Law and Policy

Edited by

Katharina Kummer Peiry

Owner and Principal, Kummer Ecoconsult, Switzerland

Andreas R. Ziegler

Professor, Université de Lausanne, Switzerland

Jorun Baumgartner

Research Fellow, Université de Lausanne, Switzerland

Edward Elgar
PUBLISHING

Cheltenham, UK • Northampton, MA, USA

Published by
Edward Elgar Publishing Limited
The Lypiatts
15 Lansdown Road
Cheltenham
Glos GL50 2JA
UK

Edward Elgar Publishing, Inc.
William Pratt House
9 Dewey Court
Northampton
Massachusetts 01060
USA

A catalogue record for this book
is available from the British Library

Library of Congress Control Number: 2016944283

This book is available electronically in the **Elgar**online
Law subject collection
DOI 10.4337/9781783473816

ISBN 978 1 78347 380 9 (cased)
ISBN 978 1 78347 381 6 (eBook)

Typeset by Servis Filmsetting Ltd, Stockport, Cheshire

Printed and bound by CPI Group (UK) Ltd, Croydon, CR0 4YY

Contents

Figures

Tables

Contributors

Jorun Baumgartner is currently a research fellow at the University of Lausanne in Switzerland. Admitted to the bar in Germany, she previously worked as a corporate lawyer in Germany, as a delegate and legal adviser at the International Committee of the Red Cross in Geneva and in different field missions, and as an economic policy officer at the International Investment Agreements section of UNCTAD. She holds an LL.M. in international and European economic law and a PhD in the field of international investment law (both from University of Lausanne), as well as a diploma of advanced studies in mediation (from FernUni Hagen). She has published in the fields of international investment law, international economic law, public international law and international criminal law.

Mirina Grosz is currently a postdoctoral researcher and lecturer at the University of Basel in Switzerland. She is admitted to the bar in Switzerland and was previously an associate in the dispute settlement trade and transport group of a leading law firm in Zurich, where she advised and represented clients before domestic courts and arbitral tribunals. She wrote her PhD in the field of WTO law/public international law and was a visiting scholar at the University of Cambridge and the Erik Castrén Institute of International Law and Human Rights at the University of Helsinki. She publishes in the fields of international economic law, public international law as well as constitutional and administrative law.

Tarcísio Hardman Reis works as a programme officer at the Secretariat of the Basel, Rotterdam and Stockholm Conventions. He has solid experience working with developing countries in the implementation of multilateral environmental agreements. Dr Reis is a certified project manager and obtained a PhD in international law from the University of Lausanne.

Katharina Kummer Peiry is owner and principal of the Swiss consulting firm Kummer EcoConsult. A specialist in international environmental law and a globally recognized expert on waste management policy, she has over 25 years of professional experience, including five years as the Executive Secretary of the Basel Convention on hazardous wastes. She taught at the University of Berne (Switzerland) for nine years, and was a Visiting Fellow at the University of New South Wales (Australia). She holds a PhD in International Law (London), a Master's of Law (Zurich), and a Certificate of Advanced Studies in Mediation (St Gallen). She has authored numerous academic publications on international environmental law and policy. 1995 saw the publication of her book *International Management of Hazardous Wastes: The Basel Convention and Related Legal Rules*, still widely considered the standard work on the Basel Convention.

Jinhui Li is currently a Professor of Tsinghua University in China and the Executive Director of Basel Convention Regional Centre for Asia and the Pacific (BCRC China). He also acts as the deputy director of the society of solid waste of Chinese Society for Environmental Sciences, a steering committee member of Regional Focal Point in Northeast Asia for Solving the E-waste Problem (StEP), a member of the Subsidiary Experts of Asia 3R Forum, and PCBs Elimination Network (PEN) Advisory Committee under the Stockholm Convention on Persistent Organic Pollutants. He has published 24 books and 237 full papers on e-waste, POPs, solid waste, hazardous waste, contaminated soil, and environmental risk assessment. He has hosted and participated in drafting and developing more than ten national waste management policy and regulations. He actively promotes technology research and development on waste recycling and pollution control, especially on e-waste.

Jessica North has worked in the waste industry for over 15 years, holding management, research, and consulting positions with a range of organizations around the world, in both developed and developing nations. Her professional experience includes landfill leachate analysis, waste options assessments, procurement, strategy development, technical due diligence, landfill emissions modelling and authoring the United Nations Environment Program's report and strategy on waste and climate change (2010). Since 2012, Dr North has been a director and manager of specialist Australian landfill gas extraction company, Landfill Gas Industries Pty Ltd (LGI). She continues her involvement in national and international waste projects.

Pierre Portas, Swiss and French national, completed academic studies on biology both in Geneva, Switzerland and San Diego, California. He spent 13 years (1976–88) at World Wide Fund for Nature (WWF-International) and the World Conservation Union (IUCN). He then joined the United Nations Environment Programme in 1989 to put in place the newly established interim secretariat for the Basel Convention. Pierre spent 16 years with the Secretariat of the Basel Convention and left in 2007 as

Deputy Executive Secretary. He also held a teaching position at the former University Institute of Development Studies in Geneva and was involved in teaching courses in many institutions around the world. He contributed to numerous scientific and technical papers and publications concerning or related to environmental issues. In 2007, he founded the NGO WE 2C (Waste Environment Cooperation Centre) based in France.

Rosemary Rayfuse is a Professor of International Law in the Faculty of Law at University of New South Wales (UNSW) in Australia and holds a conjoint appointment as Professor of International Environmental Law in the Faculty of Law, Lund University and a visiting appointment at the University of Gothenburg, both in Sweden. She is a member of the IUCN Commission on Environmental Law and the ILA Committee on Sea Level Rise and International Law. Her research focuses primarily on protection of the marine environment in areas beyond national jurisdiction and on the normative effects of climate change adaptation and mitigation responses, including climate engineering, on international law. She holds a PhD from the University of Utrecht and a Doctor of Laws honoris causa from Lund University.

Mathias Schluep studied Environmental Engineering and received his PhD in Natural Sciences from the Swiss Federal Institute of Technology in Switzerland in 2000. Dr Schluep has worked in the academic and private sector in the fields of development cooperation, environmental research and general business consultancy at national and international levels for several years. From 2006 to 2013, he worked as a program manager and senior scientist at Empa (Swiss Federal Laboratories for Materials Testing and Research). Dr Schluep has provided his expertise in sustainability issues for several cooperation projects with developing countries in Africa, Asia and Latin America for national and international government organizations and the private industry. Dr Schluep joined the World Resources Forum (WRF) in 2014 as a program director. He is leading the WRF's program activities related to mining and secondary raw materials. He currently coordinates recycling and e-waste management projects in cooperation with the Swiss State Secretariat of Economic Affairs (SECO) and the United Nations Industrial Development Organization (UNIDO).

Xiaofei Sun is currently a Senior Engineer at Tsinghua University in China and Senior Program Officer of Basel Convention Regional Centre for Asia and the Pacific (BCRC China). She also acts as the Deputy Secretary-General of Environmental Management Professional Committee, Society of Management Science of China, a member of reference group of national waste management strategy of UNEP, a member of drafting committee of the sixth 3R forum in Asia and the Pacific hosted by UNCRD. Her research areas cover the recycling and disposal technology of hazardous waste, waste water treatment, new environmental material development, and the policy on and management of solid waste with an emphasis on urban mining and renewable resources. She has hosted and participated in several national and enterprise cooperation projects.

Juliette Voïnov Kohler is Policy and Legal Adviser in the Secretariat of the Basel, Rotterdam and Stockholm Conventions, United Nations Environment Programme (UNEP). She joined the Secretariat in December 2008 and is currently the head of the Secretariat's legal team. She also teaches international environmental law at the University of Lausanne, Switzerland. Dr Voïnov Kohler has studied law in Switzerland, London and New York. She holds a PhD in international environmental law, an LLM in international law and a law degree. She was admitted to the bar in Geneva, Switzerland. Prior to joining the Secretariat, Dr Voïnov Kohler accumulated experience in international environmental law, international criminal law humanitarian issues as well as Swiss law in Switzerland (law firm, Federal Department of Foreign Affairs, Global Humanitarian Forum), New York (consultant) and Chicago (Responsibility to Protect Coalition).

Vera Weick is Programme Officer in the United Nations Environment Programme (UNEP) working in the Economics and Trade Branch in Geneva on Green Economy and related partnership initiatives. She has worked in different roles in UNEP, on economic instruments, trade and biodiversity and sustainability assessments and related capacity building initiatives and advisory services to countries. She contributed to the production of the UNEP Green Economy Report. Before joining UNEP, she worked with the German International Cooperation in Indonesia on cleaner production in enterprises and supported the development and delivery of training programmes on environmental management in South Africa, Thailand, Algeria and Vietnam. She holds a Master's in economics and business administration and a postgraduate degree from the German Development Institute.

Baoli Zhu is currently an Assistant Engineer of Tsinghua University in China and a Technical Assistant of Basel Convention Regional Centre for Asia and the Pacific (BCRC China). She mainly works on the management policy and technology research of solid waste, hazardous waste and urban mining. She has participated in several enterprise cooperation projects, and has published four full papers.

Andreas R. Ziegler is currently a Professor of International Law and the Director of the LLM Program in International and European Economic and Commercial Law at the University of Lausanne in Switzerland. Previously he was a civil servant working for several Swiss Ministries and international organizations. He has published widely on European law, public international law, on international courts and tribunals, as well as trade and investment. He regularly advises governments, international organizations, NGOs and private clients and has represented them before various domestic and international courts and arbitral tribunals. He is counsel with a law firm specialized in economic and business law (Blum & Grob Attorneys-at-law, Zurich) and on the permanent roster of panelists of the WTO and ICSID.

Foreword

When UNEP launched the Green Economy Initiative in 2008, it did so out of the conviction that without a fundamental economic transformation, the goal of sustainable development will remain elusive. UNEP's Green Economy Report, published in 2011, demonstrated that investing in environmentally significant economic sectors is not only good for the environment but also, importantly, for economic growth, jobs and social development, compared to a 'business as usual' approach.

We all recognize, however, that despite this growing engagement with green initiatives, a number of major challenges still loom, such as ecological constraints, resource availability, economic and social inequality, environment-related ill health, and persistent unemployment. Growing global and local ecological constraints are compounded by a combination of economic crises, natural disasters, and social conflict. A stronger policy strategy is required to move economic systems beyond initial investments in key sectors into the development of an inclusive Green Economy – one that prioritizes jobs, innovations, research and development, and social equity, mindful of the ecological and resource constraints.

This book takes a closer look at an area that does not immediately spring to mind when we think about a Green Economy, namely waste management, but that in fact is critical to managing both circular flow and potential environmental risks and liabilities that an economy can generate. Until recently, waste was viewed as an unwanted by-product of consumption or production, a problem rather than a resource, and something best kept out of sight and out of mind. As a result, waste management often figured at the bottom of the political agenda at all levels, and in many countries was left to municipal authorities to manage on what was primarily an ad-hoc basis. 'Reduce, Reuse, Recycle' became the message as authorities worked to balance both the problem and the potential of waste management.

With the emergence of new technologies and the use of new materials, not only have quantities of waste increased, types of waste have also become more complex and often more hazardous. Although waste reduction remains the goal, and important gains have been made in resource efficiency, waste is also being seen more and more as a potentially valuable resource for recovery and recycling of materials and energy, with significant implications for the global economy. This was formally recognized in 2011 by the governing body of the Basel Convention, the global treaty on waste management. Under the more familiar concept of 'circular economy', the practice of 'urban mining' – the extraction of precious materials from urban wastes – is a prominent example.

This book explores the hypothesis that turning wastes into valuable resources or energy might become a key area for greening the economy in a cost-effective and inclusive manner: industry could make a profit from environmentally sound resource and energy recovery from waste, provided that policies and laws at all levels facilitate the necessary operations while providing safeguards against abuse. Such an approach could also provide an incentive to invest in these operations, and thus to create decent and green jobs while protecting the environment, human health and livelihood.

Through contributions from legal, economic and technical experts in the field, the book offers an interesting range of perspectives on a key question: can waste be turned from a problem into an opportunity, and thus contribute to greening the economy? The analysis includes an assessment and experiences from Asia, a part of the world where wastes pose the greatest challenges but may also present the greatest opportunities in the future.

I would like to congratulate the authors for their efforts in bringing this important contribution to the discussion of managing waste as a scarce resource, strengthening the links that hold together a green and circular economy. By viewing the issue from the angles of law and policy, but also presenting opportunities and challenges of concrete methods and technologies, this edition will make a valuable contribution to our evolving views on waste, and the many-faceted roles it can play in advancing environmental sustainability.

Achim Steiner
Former Executive Director of UNEP and Under-Secretary-General of
the United Nations

Introduction

Andreas R. Ziegler, Katharina Kummer Peiry and Jorun Baumgartner

---

Traditionally, economic development and environmental protection were seen as opposites. The academic discussion and the relevant policy debates in the 1980s and 1990s focused on which should take priority over the other, with environmentalists and economists opposing each other over this question.

In the 1990s, the concept of Sustainable Development emerged as an attempt to achieve a balance between environmental protection and social and economic development. More recently, a number of concepts that aim at achieving a win-win situation between economic and environmental approaches and priorities have emerged. Over the last years, the international community has increasingly turned to the concept of a Green Economy with the aim of bringing the (seemingly) opposing values of economic development and environmental protection into a balance. In the absence of an internationally agreed definition, UNEP's 2011 report 'Towards a Green Economy' defines Green Economy as 'an economy that results in improved human well-being and social equity, while significantly reducing environmental risks and ecological scarcities.'[1] In operational terms, a Green Economy is an economy that promotes investments in technologies that reduce carbon emissions and pollution, enhance energy and resource efficiency or prevent the loss of biodiversity. Interrelated concepts, such as 'Green Growth', 'Green Taxation', 'Green Industry', 'Green Jobs', 'Green Accounting', have developed within the frame of specific areas and have often broadened the concept of Green Economy for their specific purposes, all the while remaining based on it.

Today the belief is consolidating that long-term sustainable development is only possible if economic development is embedded in sound

---

[1] UNEP, 'Towards a Green Economy: Pathways to Sustainable Development and Poverty Eradication', Synthesis Report (UNEP 2011) 2, available at <www.unep.org/greeneconomy/Portals/88/documents/ger/GER_synthesis_en.pdf> (last accessed on 6 December 2015).

1

environmental and social policies. What is more, States and other (both public and private) stakeholders have begun to recognize the economic potential that innovations attending to environmental and social sustainability may have in the long term. The concept of Green Economy has thus stopped being a purely 'visionary' concept and has evolved into one that is starting to find its way into concrete policy frameworks, most recently the 2015 Sustainable Development Goals adopted by the UN General Assembly.[2] Nevertheless, some scepticism remains, as seen for example in the negotiations at the Rio+20 Summit in 2012: developing countries in particular were wary of supporting this concept, which they feared might simply be a new way of erecting trade barriers and slowing economic growth in the developing world.[3]

Much of the discussion and effort related to achieving a Green Economy have thus far focused on the areas of climate change and energy, with other areas of environmental protection, including waste management, receiving limited attention. Waste, subject to environmental legislation in the developed world since the 1970s, has until recently been seen as an unwelcome and costly by-product of modern societies, and thus as a problem. Accordingly, waste legislation at all levels has long focused on final disposal, and since the 1980s, on controlling export and import and preventing illegal traffic in waste, especially from developed to developing countries.[4] The year 1989 saw the adoption of the Basel Convention on the Control of Transboundary Movements of Hazardous Wastes and their Disposal, negotiated under the auspices of UNEP to protect developing countries from being used as cheap disposal grounds for hazardous wastes from industrialized countries. However, the implementation of the Convention, and waste management in general, have remained at the bottom of the political agenda at all levels. Funding is still largely insufficient to ensure environmentally sound management of wastes, especially in the developing world.

---

[2] See UN, 'Transforming our World: The 2030 Agenda for Sustainable Development', General Assembly Resolution 70/1 (25 September 2015) available at <www.un.org/ga/search/view_doc.asp?symbol=A/RES/70/1&Lang=E> (last accessed on 24 February 2016). See also e.g. UNCTAD, 'World Investment Report (WIR) 2014 – Investing in the SDGs: An Action Plan' (UNCTAD 2014) available at <http://unctad.org/en/PublicationsLibrary/wir2014_en.pdf> (last accessed on 24 February 2016).

[3] See K. Kummer, R. Khanna and V. Sahajwalla, 'Resource and Energy Recovery from Wastes: Perspectives for a Green Economy' (2012) 42(6) *Environmental Policy and Law* 344.

[4] For an overview see Katharina Kummer, *International Management of Hazardous Wastes, the Basel Convention and Related Legal Rules* (Oxford University Press 1995, reprinted 1999).

And yet, waste management remains one of the great challenges of our times. With a world population estimated to reach over nine billion people by 2050, resource consumption will continue to skyrocket, leading to the generation of dramatic dimensions of waste. The increase in available income in developing and emerging countries will further accelerate this waste generation.[5] The massive increase in waste raises a host of problems that may ultimately touch upon humans' very basis of existence. If not properly handled, wastes may jeopardize human livelihood either directly (e.g. through contamination) or indirectly (e.g. through its impact on climate change). The more waste is generated in the world, the more urgent the problem thus becomes if adequate solutions are not found.

In line with continuing efforts to promote sustainable development in recent years, efforts have been made to prevent waste generation and increase recycling rates. New technologies have evolved that make this possible. New waste streams have emerged over the last decades, including in particular waste electrical and electronic equipment (WEEE), currently one of the fastest growing waste streams worldwide. With the progress of globalization, waste management is no longer a problem of individual nations but one that has attained a global dimension. At the same time, some waste materials, including certain components of WEEE, are also secondary resources for which a market exists.[6] Resource recovery from waste is in some cases less energy intensive than primary production, and energy recovery can reduce primary energy consumption. Yet many challenges remain, both in legal and practical terms. One of these is the widespread illegal traffic in certain types of wastes, in particular WEEE, for improper recycling,[7]

---

[5] UNEP, 'Towards a Green Economy (n 1), 17–18: '. . . [a]s living standards and incomes rise, the world is expected to generate over 13.1 billion tons of waste in 2050, about 20% higher than the amount in 2009', available at <www.unep.org/greeneconomy/Portals/88/documents/ger/GER_synthesis_en.pdf> (last accessed on 2 September 2015).

[6] A recent report on illegal WEEE trade estimated that '. . . the value of recycling of WEEE will be €2.15–3.67 bn by 2020', see J. Huisman, I. Botezatu, L. Herreras et al, 'Countering WEEE Illegal Trade (CWIT) Summary Report, Market Assessment, Legal Analysis, Crime Analysis and Recommendations Roadmap' (30 August 2015) (Lyon, France) 18, available at <www.cwitproject.eu/wp-content/uploads/2015/08/CWIT-Final-Summary1.pdf> (last accessed on 6 September 2015).

[7] 'CWIT Summary Report' (n 6) 6: '. . . in Europe, only 35% (3.3 million tons) of all WEEE discarded in 2012 ended up in the officially reported amounts of collection and recycling systems. The other 65% (6.15 million tons) was either exported (1.5 million tons), recycled under non-compliant conditions in Europe

despite the enactment of legislation aimed at managing and controlling the relevant trade.[8]

For waste management, a Green Economy approach would mean making the so-called 'life-cycle approach' work within the broader goal of economic development, creating economic opportunities within the policy goals of reduction, reuse or recycling of wastes. The potential that lies in a resource-based approach towards waste management has started to transpire with the shift in focus by the international community to the overarching policy goal of sustainable development. The Basel Convention features provisions mandating waste minimization and environmentally sound waste management policies. In 1999, the Fifth Conference of the Parties, on the occasion of the tenth anniversary of the Convention, made efforts in these areas a priority for the following decade. The Tenth Conference of the Parties in 2011 acknowledged that wastes that cannot be prevented can be valuable resources, and supported the concept of waste prevention and waste management as a legitimate economic opportunity. In a similar vein, the Rio+20 Summit in June 2012 highlighted a Green Economy as a possible tool for promoting sustainable development, and called for reduction, reuse and recycling of waste, recognizing the need for public-private partnerships in these areas. The Sustainable Development Goals (SDGs), which replace the Millennium Development Goals (MDGs), touch in many of the 17 overarching goals either directly or indirectly on how wastes and their management can contribute to sustainable development.[9]

International transfer of waste also features in the discussion on trade and environment, including in the framework of the World Trade Organization. Waste and trade do have a peculiar relationship, and globalization has its own role to play. Waste is both a consequence of globalized trade as it may be one of its goods. Regarding waste from the trade angle does have the advantage of perceiving the monetary value of waste and its management, from collection to recycling.[10] It also has the

---

(3.15 million tons), scavenged for valuable parts (750,000 tons) or simply thrown in waste bins (750,000 tons)'.

[8]   See e.g. European Union, Directive 2012/19/EU of the European Parliament and of the Council of 4 July 2012 on waste electrical and electronic equipment (WEEE) (2012) OJ L 197, 38–71.

[9]   See in particular SGDs 6.3, 7.1, 7.2, 7.a, 8, 11.6, 12.4 and 12.5, <https://sustainabledevelopment.un.org/sdgs> (last accessed on 24 February 2016).

[10]   UNEP estimated the value of the world market for waste, from collection to recycling, to be worth around USD 40 billion a year, see UNEP, 'Towards a Green Economy' (n 1) 18.

merit of shifting the focus from perceiving waste (only) as a problem to perceiving it (also) as a potential resource, in other words, an opportunity. It would appear, then, that waste management deserves more attention in the discussion of a Green Economy and related concepts than it is currently receiving.

The key question that inspired this volume is whether waste management has the potential to become a pilot area of a Green Economy, and if so, what would be required to achieve this. Might waste management, being a less controversial and less complex issue than climate change – often portrayed as the main driver of a Green Economy – have the technical, economic and social potential, as yet insufficiently explored, to move this concept forward? The potential implications appear attractive: can industry make a profit from the relevant operations if the applicable legal and policy frameworks facilitate the necessary operations while providing safeguards against abuse? Can this in turn serve as an incentive for industry to invest in these operations, and create green business opportunities and green jobs while protecting the environment, and human health and livelihood? Might this approach contribute to addressing the problem of illegal trade and improper recycling of hazardous wastes by making the legitimate alternatives more attractive?

There is no scarcity in literature on international environmental law in general[11] and the international law perspective on selected environmental problems.[12] Much has also been written about the meaning and

---

[11] See among many others: Alexandre Kiss and Dinah Shelton, *International Environmental Law* (Transnational Publishers 1991); Patricia Birnie and Alan Boyle, *International Law and the Environment* (Oxford University Press 1994); David Hunter, James Salzmann and Durwood Zaelke, *International Environmental Law* (Routledge 1998); Edith Brown Weiss, *International Environmental Law and Policy* (Aspen Publishers 2006); Malgosia Fitzmaurice, David Ong and Panos Merkouris (eds), *Research Handbook of International Environmental Law* (Edward Elgar 2010); Ulrich Beyerlin and Thilo Marauhn, *International Environmental Law* (Hart Publishing 2011); Paul Harris and Graeme Lang (eds), *Routledge Handbook of Environment and Society in Asia* (Routledge 2015).

[12] See among many others: S. Jayakumar, Tommy Koh, Robert Beckman and Hao Duy Phan (eds), *Transboundary Pollution, Evolving Issues of International Law and Policy* (Edward Elgar 2015); Prati Pal Singh and Vinod Sharma (eds), *Water and Health* (Springer 2014); Willem Wijnstekers, *The Evolution of CITES* (International Council for Game and Wildlife Conservation, 9th ed., 2011); Rosemary Rayfuse and Shirley Scott (eds), *International Law in the Era of Climate Change* (Edward Elgar 2012); Frank Maes et al (eds), *Biodiversity and Climate Change* (Edward Elgar 2015); C. Bail, R. Falkner, H. Marquard (eds), *The Cartagena Protocol on Biosafety: Reconciling Trade in Biotechnology with*

implementation of the concept of sustainable development,[13] and the notion of Green Economy has in recent times received almost as much attention.[14] By contrast, there are relatively few scholarly contributions on the specific topic of waste management,[15] and even less that explore more in depth the role the concept of Green Economy could or should play in the environmentally sound handling of wastes. The objective of this book is to attempt such an analysis.

The book starts out with an exploration of the issues from a legal and policy angle: the first part sets the scene for exploring the international legal framework (in particular international environmental law, international trade law and human rights law) and its gaps. The law, however, does not exist in a void, but has been developed to deal with the facts of waste and materials management. The second part therefore delves into

---

Environment and Development? (The Royal Institute of International Affairs/ Earthscan 2002).

[13]   See e.g. Giles Atkinson, Simon Dietz, Eric Neumayer and Matthew Agarwala (eds), *Handbook of Sustainable Development* (Edward Elgar, 2nd ed. 2014); Malgosia Fitzmaurice, Sandrine Maljean-Dubois and Stefania Negri (eds), *Environmental Protection and Sustainable Development from Rio to Rio+20* (Brill/ Nijhoff 2014).

[14]   See e.g. Adrian Newton and Elena Cantarello (eds), *An Introduction to the Green Economy: Science, Systems and Sustainability* (Routledge/Earthscan 2014); Robert Richardson (ed.), *Building a Green Economy, Perspectives from Ecological Economics* (Michigan State University Press 2013); UNEP, *Green Economy and Trade: Trends, Challenges and Opportunities* (UNEP 2013); Anneleen Kenis and Matthias Lievens (eds), *The Limits of the Green Economy, From Reinventing Capitalism to Repoliticising the Present* (Routledge 2012); Elena Merino-Blanco, *Natural Resources and the Green Economy* (Martinus Nijhoff Publishers 2012); Dan Brockington, 'A Radically Conservative Vision?' (2012) 43 *Development and Change* 409; José A. Puppim de Oliveira (ed.), *Green Economy and Good Governance for Sustainable Development: Opportunities, Promises and Concerns* (United Nations University Press 2012); David Pearce, Anil Markandya and Edward B. Barbier, *Blueprint for a Green Economy* (Earthscan Publiations 1990).

[15]   See e.g. Thomas Kinnaman and Kenji Takeuchi (eds), *Handbook on Waste Management* (Edward Elgar 2014); Michikazu Kojima and Etsuyo Michida (eds), *International Trade in Recyclable and Hazardous Waste in Asia* (Edward Elgar 2013); Syeda Azeem Unnisa and Bhupatthi S. Rav (eds), *Sustainable Solid Waste Management* (Apple Academic Press 2013); Mirina Grosz, *Sustainable Waste Management under WTO Law* (Brill/Nijhoff 2011); Trevor Letcher and Daniel Vallero (eds), *Waste: A Handbook for Management* (Academic Press 2011); Katharina Kummer, *International Management of Hazardous Wastes, the Basel Convention and Related Legal Rules* (Oxford University Press 1995, reprinted 1999); Jonathan Kruger, *International Trade and the Basel Convention* (The Royal Institute of International Affairs/Earthscan Publications 1999).

different economic and technical issues of waste management that afford a glimpse of aspects that go beyond the law.

The book brings together contributions from an interdisciplinary group of authors who have made significant contributions to relevant academic and policy discussions and publications in their respective fields. It attempts to add to the academic analysis a consideration of concrete results on the ground, thus offering academic as well as practical perspectives on the questions.

PART I

Resource and energy recovery from wastes in international environmental law and policy

1. Principles of international environmental law applicable to waste management

Rosemary Rayfuse

EXECUTIVE SUMMARY

General principles of international environmental law provide the theoretical foundation for the development of normative frameworks in international law. In the waste management context, five general principles are particularly relevant: the principle of permanent sovereignty over natural resources and the duty not to cause transboundary harm; the principle of preventive action; the corresponding principle of cooperation; the principle of sustainable development; and the precautionary principle. Operationalization of these principles in the waste context has led to the development of new principles, such as those of self-sufficiency, proximity, waste minimization, environmentally sound management and prior informed consent, all of which are further operationalized in the detailed rules set out in the Basel Convention and other treaties dealing with waste management. This chapter examines the interpretation and application of these general principles and the role they have played in the development of the international legal regime for the management and transboundary movement of waste.

1.1 INTRODUCTION

'One person's waste is another person's treasure' – or so the saying goes. But treasures can be an impossible burden, particularly where adequate resources, facilities and capacity are not available for their care, control, management and maintenance. In our increasingly disposable consumer society, our wasteful treasure threatens to overwhelm us in both volume and toxicity.[1] Its control and management is therefore of vital importance for humankind.

While primarily a matter of domestic concern, the emergence of economic incentives for States to dispose of waste in other States has turned the issue of waste management into one of international concern. Of particular disquiet has been the propensity towards 'toxic colonialism', or the practice of developed States exporting their waste to developing States less able to deal with it.[2] The increasing awareness of potential harm from mismanagement and disposal of waste, together with its global regulation, have reduced the incidence of dumping of waste by developed States into developing States, although the practice remains a concern, with estimates reportedly showing more than 50 per cent of worldwide transboundary waste movements as illegal.[3] More recently, however, the concept of waste as a potentially valuable resource has started to take hold with developing States increasingly seeking to import waste, in particular e-waste, for the economic opportunities its recycling, and the recovery of the precious metals used in its production, present.[4] The question thus arises as to the nature and content of international law relating to waste management, and its efficacy in addressing the dangers posed by poorly regulated transboundary movement of wastes. While subsequent chapters in this volume discuss the relevant rules of international law applicable to waste management in detail, this chapter explores the general principles of international environmental law relevant to the management and transboundary movement of waste. Before doing so, however, some preliminary comments on the role of general principles are warranted.

---

[1]   Secretariat of the Basel Convention, 'Wastes without Frontiers: Global Trends in Generation and Transboundary Movements of Hazardous Wastes and other Wastes' (2011) available at <http://archive.basel.int/pub/ww-frontiers31Jan2010.pdf> (last accessed on 14 August 2015).

[2]   Jennifer R. Kitt, 'Waste Exports to the Developing World: A Global Response' (1995) 7 *Georgetown International Environmental Law Review* 485.

[3]   Katharina Kummer Peiry, 'Basel Convention: Turning Wastes into Valuable Resources – Promoting Compliance with Obligations?' (2011) 41(4/5) *Environmental Policy and Law* 177, 178.

[4]   O. Osibanjo and I.C. Nnorom, 'The Challenge of Electronic Waste (e-waste) Management in Developing Countries' (2007) 25 *Waste Management and Research* 489.

1.2   THE ROLE OF PRINCIPLES IN INTERNATIONAL ENVIRONMENTAL LAW

Like international law in general, international law relating to waste management is not simply a mosaic of specific rules adopted in treaties. Rather, it can more appropriately be described as a system governing the international relations among States and other entities in respect of their activities relating to waste management and, in particular, the transboundary movement of waste. This system consists of both specific treaty-based rules and rules of customary international law as well as general principles. The emphasis here on principles is deliberate. Admittedly, the content and legal status of principles is less clear than that of binding rules, and their invocation, unlike that of rules, does not lead inexorably to any particular decision. As Dworkin puts it:

> [A]ll that is meant, when we say that a particular principle is a principle of our law, is that the principle is one which officials must take into account, if it is relevant, as a consideration inclining in one way or another.[5]

However, principles play a valuable role in integrating various legal, economic, social and political considerations into various fields of international law.[6] They also provide guidance on the interpretation and application of relevant rules in situations of conflicting interpretation.[7] In addition, they provide predictable parameters for environmental protection and can provide the orientation for the development of the law.[8] Thus, principles serve as the theoretical basis for the rules we adopt and the framework within which those rules are to be applied.[9]

To fully understand and assess the operation and efficacy of the rules

---

[5]   Ronald Dworkin, *Taking Rights Seriously* (Harvard University Press 1977) 24, 26.

[6]   Patricia Birnie, Alan Boyle and Catherine Redgwell, *International Law and the Environment* (Oxford University Press, 3rd ed. 2009) 109.

[7]   Ibid.

[8]   See, e.g., Lluis Paradell-Truis, 'Principles of International Environmental Law: An Overview' (2000) 9 *Review of European Community and International Environmental Law* 93, 95–7; Laurence Boisson de Chazournes, 'Features and Trends in International Environmental Law' in Yann Kerbrat and Sandrine Maljean-Dubois, *The Transformation of International Environmental Law* (Pedone and Hart 2011) 11.

[9]   *Gentini (Italy v Venezuela)* (1913) 10 RIAA 551, cited in Philippe Sands and Jacqueline Peel, *Principles of International Environmental Law* (Cambridge University Press, 3rd ed. 2012) 189.

of international law applicable to waste management it is first necessary to develop an appreciation of the cardinal principles of international law that are applicable in this context. This may include principles emanating from a number of areas of international law dealing, inter alia, with sustainable development, human rights, international watercourses, law of the sea, armed conflict or international trade and include the more general principles relating to state responsibility. However, for present purposes, this chapter focuses on the relevant principles of that body of law known as international environmental law.

A glance at any international environmental law text will reveal a plethora of principles, some contested, some well accepted, that are applicable to various environmental issues. While there is no single agreed taxonomy of environmental law principles, the following (non-exhaustive) list of general principles can be identified as most relevant to the current enquiry:

- the principle of permanent sovereignty over natural resources and the duty not to cause transboundary harm;
- the principle of preventive action;
- the principle of cooperation;
- the principle of sustainable development; and
- the precautionary principle.

In the waste management context, these general principles are supplemented by other principles such as those set out in the 1987 Cairo Guidelines and Principles for the Environmentally Sound Management of Hazardous Wastes,[10] which sets out 29 principles designed to assist governments to develop policies for environmentally sound management of hazardous wastes from generation to final disposal, all of which essentially derive from, and seek to operationalize, the more general principles referred to above. Many of the Cairo Guidelines and Principles have been incorporated into the regimes established by the 1989 Basel Convention on the Control of Transboundary Movements of Hazardous Wastes and their Disposal[11] (Basel Convention) and other regional conventions dealing with the subject.

Of particular relevance are:

---

[10]   UNEP/GC.14/17 (1987) Annex II, approved by UNEP/GC/Dec./14/30 (1987) UNEP ELPG No. 8.
[11]   Adopted on 22 March 1989, entered into force on 24 May 1992 (1989) 28 ILM 657.

- the principle of self-sufficiency;
- the principle of proximity;
- the principle of minimization of waste;
- the principle of environmentally sound management; and
- the principle of prior informed consent.

It is important to note that not all of these principles enjoy the same binding legal status. Some principles, such as the principle of permanent sovereignty over natural resources, the no-harm principle and the principles of preventive action and cooperation, are accepted as enjoying the status of customary international law and are hence binding on all States. The binding status of the precautionary principle, however, remains contested,[12] although both the International Court of Justice (ICJ)[13] and the Seabed Disputes Chamber of the International Tribunal for the Law of the Sea (ITLOS) have recognized a 'trend towards making [the precautionary approach] part of customary international law'.[14] In the case of sustainable development, its very legal nature as a principle is contested, even though its force and imperative as a 'concept' or a 'goal' is accepted.[15] The customary status of the remaining principles is also open to debate. Thus, while they may now be binding in the waste management context as a matter of treaty law, questions remain as to their binding

---

[12] For opinions in support see, e.g., James Cameron and Juli Abouchar, 'The Status of the Precautionary Principle in International Law' in David Freestone and Ellen Hey (eds), *The Precautionary Principle and International Law: The Challenge of Implementation* (Kluwer Law International 1996) 29, 29–52; Arie Trouwborst, *Evolution and Status of the Precautionary Principle in International Law* (Kluwer Law International 2002) 286; Philippe Sands, *Principles of International Environmental Law* (Cambridge University Press, 2nd ed. 2003) 279; Owen McIntyre and Thomas Mosedale, 'The Precautionary Principle as a Norm of Customary International Law' (1997) 9 *Journal of Environmental Law* 221, 235. For more cautious assessments see, e.g., Pascale Martin-Bidou, 'Le principe de précaution en droit international de l'environnement' (1999) 103 *Revue générale de droit international public* 631, 658–65 and Daniel Bodansky, 'Remarks' (1991) 85 *Proceedings of the American Society of International Law* 413.

[13] See *Pulp Mills on the River Uruguay (Argentina v Uruguay)* (Judgment) (2010) ICJ Rep 2010, 14 (para. 164).

[14] *Responsibilities and Obligations of States Sponsoring Persons and Entities with Respect to Activities in the Area*, ITLOS Case No. 17, Advisory Opinion (1 February 2011) para. 135.

[15] Vaughan Lowe, 'Sustainable Development and Unsustainable Arguments' in Alan Boyle and David Freestone (eds), *International Law and Sustainable Development: Past Achievements and Future Challenges* (Oxford University Press 1999) 19–38.

nature vis-à-vis non-parties. Their importance, however, cannot be overstated and they are thus examined here within the context of a discussion of the general international environmental law principles identified above and their application in the international waste management context.

1.3 PERMANENT SOVEREIGNTY OVER NATURAL RESOURCES AND THE 'NO-HARM' PRINCIPLE

The principle of permanent sovereignty over natural resources has its origins in the various resolutions adopted by the United Nations General Assembly beginning in the early 1950s.[16] Initially intended to balance the rights of States over their resources with the desire of foreign companies for legal certainty in respect of their investments, the principle was formulated in terms that allowed States to conduct or authorize such activities as they may choose within their territories, subject only to any limitations established under international law. By the 1970s, States recognized that limitations to the application of the principle were necessary, particularly in order to protect the environment. Thus, while Principle 21 of the 1972 Stockholm Declaration[17] affirms the sovereign right of States to exploit their own resources as they see fit, it conditions this sovereignty by imposing on States 'the responsibility to ensure that activities within their jurisdiction or control do not cause damage to the environment of other States or to areas beyond the limits of national jurisdiction'. This so-called 'no-harm principle' was first articulated in the Trail Smelter arbitration[18] where its application was originally only discussed in the context of transboundary harm to other States. An important aspect of the formulation in Principle 21 is that the principle also now applies to areas beyond national jurisdiction, thereby providing the foundation for the various prohibitions or restrictions on the dumping of wastes and other matter on and into the high seas,[19]

---

[16] See, e.g., UNGA Res. 525(VI) (1950); Res. 626(VII) (1952); Res. 837(IX) (1954); Res. 1314 (XIII) (1958); Res. 1515(XV) (1960).
[17] Stockholm Declaration of the United Nations Conference on the Human Environment, 16 June 1972 (1972) 11 ILM 1416.
[18] *Trail Smelter (USA v Canada)* (1941) 3 RIAA 1905, 1965.
[19] See, e.g., the 1972 Convention on the Prevention of Marine Pollution by Dumping of Wastes and Other Matter (adopted on 29 December 1972, entered into force on 30 August 1975) 1046 UNTS 138; and its 1996 Protocol to the Convention on the Prevention of Marine Pollution by Dumping of Wastes and Other Matter (adopted on 7 November 1996, entered into force on 24 March 2006) (1997) 36 ILM 1.

in Antarctica,[20] into the atmosphere,[21] and into rivers and other freshwater bodies.[22] So accepted is the language of Principle 21 that it was reiterated verbatim in Principle 2 of the 1992 Rio Declaration,[23] and the customary status of the combined 'permanent sovereignty/no-harm' principle was confirmed by the ICJ in its 1996 Advisory Opinion on the *Legality of the Threat or Use of Nuclear Weapons*.[24]

The principle of permanent sovereignty acts as a double-edged sword. States have the freedom to exploit their resources and reap the benefits therefrom. They are also entitled to be free from interference by other States. Thus, the no-harm principle operates to constrain the activities of States where the potential for transboundary harm exists, although the principle does not answer the questions as to what constitutes environmental damage, what level of damage or harm is prohibited, whether the obligation is one of absolute liability, strict liability, or fault-based liability, what the consequences of a violation might be or the extent of any liability. The answers to these questions must thus be found in treaties and in State practice.

In the waste management context the application of the permanent sovereignty/no-harm principle means that States are free to generate waste, but they must not dispose of it in a manner that causes harm to the environment of other States or to areas beyond national jurisdiction. This tension between the dictates of sovereignty and the recognition of the potential for transboundary impacts of waste disposal lies at the very heart of the international regime established by the Basel Convention and

---

[20]   Art. 4 (6) of the Basel Convention on the Control of Transboundary Movements of Hazardous Wastes and Their Disposal (Basel Convention) (adopted on 22 March 1989, entered into force on 5 May 1992) 1673 UNTS 57.
[21]   See, e.g., the 1979 Convention on Long-range Transboundary Air Pollution (adopted on 13 November 1979, entered into force on 16 March 1983) 1302 UNTS 217; the 1987 Montreal Protocol on Substances that deplete the Ozone Layer (adopted on 16 September 1987, entered into force on 1 January 1989) 1522 UNTS 3; and the 1997 Kyoto Protocol to the United Nations Framework Convention on Climate Change (adopted on 11 December 1997, entered into force on 16 February 2005) 2303 UNTS 148, which prohibit or regulate the dumping of certain noxious gases into the atmosphere.
[22]   Convention on the Protection and Use of Transboundary Watercourses and International Lakes (adopted on 17 March 1992, entered into force on 6 October 1996) 1936 UNTS 269, Art. 2(2)(a).
[23]   Rio Declaration on Environment and Development, Report of the United Nations Conference on Environment and Development, I (1992) UN Doc A/CONF.151/26, (1992) 31 ILM 874.
[24]   *Legality of the Threat or Use of Nuclear Weapons* (Advisory Opinion) (1996) ICJ Reports 226.

by the regional conventions dealing with the subject. As the Preamble to the Basel Convention makes clear, 'all states have the sovereign right to ban the entry or disposal of foreign hazardous wastes and other wastes in their territory'. In other words, while international law places no, or at any rate few, limits on waste generation, and no outright ban on trade, all States have the sovereign right to determine whether to receive waste and, if so, what impacts on their territory they will be willing to accept. It has been noted that this represents an important gloss on the no-harm principle in that, unlike State practice in other areas such as nuclear instal-lations, air pollution, or international watercourses, where transboundary effects are permitted unless certain levels of harm occur, 'it cannot be assumed that waste disposal in other states is permitted unless it is shown to be harmful'.[25] The criteria of harm, or even potential harm, has been removed in favour of the absolute sovereignty of States to decide for them-selves, either individually or regionally, whether or not to receive waste; although, as the terminology of 'hazardous waste' implies, the criteria of harm is not wholly irrelevant.

Despite the sovereign right to refuse imports, the Basel Convention, as its full name implies, merely establishes a regime to control trade in hazard-ous waste rather than prohibit it. Encapsulated in the terminology of the principles of 'self-sufficiency', 'proximity' and 'prior informed consent', the Basel Convention requires each State to reduce its waste generation to a minimum,[26] to become self-sufficient in waste management,[27] and to dispose of waste as close as possible to the place of generation.[28] To that end, parties must ensure that adequate waste facilities are located within their jurisdiction, although this is qualified by 'to the extent possible'.[29] Flowing from the principle of State sovereignty, parties are entitled to prohibit the import of any hazardous or other wastes and must consent in writing to any specific imports they have not prohibited.[30] Needless to say, parties must not allow the export of waste to other parties who have prohibited it.[31] Parties are also required to provide information on proposed transboundary movements of hazardous and other wastes to any State concerned and they are to prevent imports if they have reason to believe that the imports will not be managed in an environmentally

---

25   Birnie, Boyle and Redgwell (n 6) 473.
26   Basel Convention (n 20) Art. 4 (2)(a).
27   Ibid., Art. 4 (9)(a).
28   Ibid., Art. 4 (2)(b).
29   Ibid.
30   Ibid., Art. 4 (1)(b).
31   Ibid., Art. 4 (1)(a).

sound manner.[32] All shipments are subject to the requirements of the prior written consent of any party through which or to whom waste is being exported.[33]

While as a matter of basic treaty law the Basel Convention is binding only on its States parties, the regime is cleverly designed to have at least some third-party effect by imposing a legal obligation on parties not to permit export to or import from non-State parties. However, recognizing that the right to accept waste imports is also a sovereign right of any State wishing to do so, this prohibition is subject to an exception where the States concerned have entered into another bilateral, multilateral or regional agreement or arrangement, provided that it does not derogate from the requirement of environmentally sound management of hazardous and other wastes found in the Basel Convention.[34] Areas beyond national jurisdiction are also protected by the prohibition on the export of wastes for disposal in the Antarctic area,[35] even if their transportation is not transboundary in nature.

Thus, while the Basel Convention seeks to discourage export of hazardous and other wastes, the possibility of transboundary shipments remains, although they must be reduced to the 'minimum consistent with environmentally sound and efficient management',[36] and should only be permitted if the State of export lacks the technical capacity and necessary facilities, capacity and suitable disposal sites to do so, or, importantly, where the waste is intended for recycling or recovery.[37] In the past these exceptions have been seriously contested by developing States concerned that the Basel regime fails to address the control of shipments of mixed waste, instances of inadequate or inappropriate disposal by importing States, and the problems of forgery, bribery and corruption circumventing the notice and consent provisions.[38] Exercising their sovereign rights States have therefore entered into other agreements more restrictive than the Basel Convention.

The 1991 Convention on the Ban of Import into Africa and the Control

---

[32] Ibid., Art. 4 (2)(f) and (g).

[33] Ibid., Art. 6.

[34] Ibid., Art. 11.

[35] Defined, consistent with the Antarctic Treaty, as south of 60 degrees south. This therefore includes both the Antarctic continent and the surrounding Southern Ocean, see Basel Convention (n 20) Art. 4 (6).

[36] Ibid., Art. 4 (2)(d).

[37] Ibid., Art. 4 (9)(a) and (b).

[38] John Ovink, 'Transboundary Movement of Hazardous Waste, the Basel and Bamako Conventions: Do Third World Countries Have a Choice?' (1995) 13 *Dickinson Journal of International Law* 281, 285.

of Transboundary Movement and Management of Hazardous Wastes within Africa (Bamako Convention),[39] prohibits outright all trade in hazardous waste and requires its parties to prohibit the import of all wastes into Africa from non-contracting parties and to deem such imports illegal and criminal.[40] Parties must ensure that any hazardous wastes to be exported are managed in an environmentally sound manner in the States of import and transit, and only authorized persons can store such wastes.[41] Importantly, even wastes to be used as raw materials for recycling and recovery may not be exported.[42] The 1995 Convention to Ban the Importation into Forum Island Countries of Hazardous and Radioactive Wastes and to Control the Transboundary Movement of Hazardous Wastes within the South Pacific Region (Waigani Convention)[43] similarly bans the import of hazardous and radioactive wastes into its area of coverage and regulates their transboundary movement between the parties.[44] In addition, two parties, Australia and New Zealand, are required to ban the export of hazardous wastes to all Forum Island countries and territories within the Convention area.[45] The 1992 Central American Agreement on Hazardous Waste[46] bans all imports of hazardous and radioactive wastes and of toxic substances not permitted in the country of manufacture, while the Barcelona Convention Waste Trade Protocol[47] prohibits the export of hazardous and radioactive wastes to non-OECD countries and parties

---

[39] Convention on the Ban of Import into Africa and the Control of Transboundary Movement and Management of Hazardous Wastes within Africa (Bamako Convention) (Adopted on 29 January 1991, entered into force on 22 April 1998) (1991) 30 ILM 775.

[40] Ibid., Art. 4 (1).

[41] Ibid., Art. 4 (3)(i) and (m)(i).

[42] Ibid., Art. 5 (4).

[43] Convention to Ban the Importation into Forum Island Countries of Hazardous and Radioactive Wastes and to Control the Transboundary Movement of Hazardous Wastes within the South Pacific Region (Waigani Convention) (adopted on 16 September 1995, entered into force 21 October 2001) 2161 UNTS 93.

[44] Ibid., Art. 4 (1).

[45] Ibid., Art. 4 (1)(b).

[46] Central American Agreement on Hazardous Waste (adopted on 11 December 1992, entered into force 17 November 1995). See <www.ecolex.org/ecolex/ledge/view/RecordDetails?id=TRE–001167&index=treaties> (last accessed on 13 August 2015).

[47] Barcelona Convention for the Protection of the Mediterranean Sea against Pollution: Protocol on the Prevention of Pollution of the Mediterranean Sea by Transboundary Movements of Hazardous Wastes and their Disposal, 1 October 1996, UNEP(OCA)/MED/IG.9/4, 11 October 1996.

that are not members of the European Community are prohibited from importing hazardous and radioactive wastes.

These efforts have been echoed in the Conference of the Parties to the Basel Convention which, in 1994, approved an immediate ban on the export from OECD countries to non-OECD countries of hazardous wastes intended for final disposal and also agreed to ban the export of wastes intended for recovery and recycling by 31 December 1997.[48] Known as the 'Basel Ban', disputes as to its legally binding nature were resolved by the adoption, the following year, of the Basel Ban Amendment to the Convention[49] which seeks to ban hazardous waste exports for both final disposal and recycling from Annex VII parties (EU, OECD and Lichtenstein) to non-Annex VII parties. The Amendment has yet to enter into force but provides further evidence, if any were needed, of the application of the permanent sovereignty and no harm principles in the international regime regulating the transboundary movement of hazardous wastes.

1.4   THE PRINCIPLE OF PREVENTIVE ACTION

Closely related to the no-harm principle, the principle of preventive action obliges States to prevent damage to the environment and to reduce, limit or control activities that might cause or risk such damage. Confirmed as a rule of customary international law by the ICJ in the *Pulp Mills* case,[50] the arbitral tribunal in the *Iron Rhine* case recognized that it is not just 'a principle of general international law' that 'applies in autonomous activities', but that it also 'applies in activities taken in implementation of specific treaties between the Parties'.[51] The obligation is not, however, absolute.[52] Rather, it is one of due diligence which, 'entails not only the adoption of appropriate rules and measures, but also a certain level of vigilance in their enforcement and the exercise of administrative control applicable to

---

[48]   Decision II/12, Report of COP2, UNEP/CHW.2/30, 25 March 1994.

[49]   Decision III/1, Report of COP3, UNEP/CHW.3/34, 17 October 1995. For discussion see Louise de la Fayette, 'Legal and Practical Implications of the Ban Amendment to the Basel Convention' (1995) 6 *Yearbook of International Environmental Law* 703.

[50]   *Pulp Mills* (n 13) para. 101.

[51]   *Iron Rhine Railway* Arbitration *(Belgium v The Netherlands)* (2005) 27 RIAA 35 (paras 59 and 222).

[52]   ILC Draft Articles on Prevention of Transboundary Harm from Hazardous Activities (2001) Art. 3. See Yearbook of the International Law Commission (2001-II) Part 2, para. 7.

public and private operators, such as the monitoring of activities under-taken by such operators'.[53]

The objective of the preventive principle is to minimize environmental damage. To that end, it requires action to be taken at an early stage, before damage has actually occurred. Importantly, the principle applies whether that damage might be transboundary or confined to areas under national jurisdiction.[54] This approach is justified on the basis that damage to the environment is often irreversible and mechanisms for reparation of environmental damage are seriously limited.[55] In this respect the principle operates as a precautionary brake on State action. However, the degree of 'due diligence' and the action to be taken will vary depending, inter alia, on the nature of the specific activities, the technical and economic capabilities of States, and the effectiveness of their territorial control.[56] In addition, 'measures considered sufficiently diligent at a certain moment may become not diligent enough in light, for instance, of new scientific or technological knowledge', and 'can change in relation to the risks involved in the activity'.[57] As such, the obligation requires States 'to take [reasonably appropriate] measures within [their] legal systems'[58] and to ensure that those measures are both effective and that they 'reflect the environmental and developmental context to which they apply'.[59] In other words, the content of due diligence is a changing one that requires States to 'move with the times'.

In the waste management context, international law has traditionally taken no, or at least little, position on the generation of waste, focus-ing rather on its disposal and transboundary movement. For example, Principle 6 of the Stockholm Declaration calls merely for a halt to the discharge, not generation, of toxic or other substances while Principle 14 of the Rio Declaration similarly calls only for effective cooperation 'to discourage or prevent the relocation or transfer to other states of any activities and substances that cause severe environmental degradation or are found to be harmful to human health'. With the exception of treaties

---

[53]   *Pulp Mills* (n 13) para. 197.
[54]   Philippe Sands and Jacqueline Peel, *Principles of International Environmental Law* (Cambridge University Press, 3rd ed. 2012) 201.
[55]   *Gabčíkovo-Nagymaros (Hungary v Slovakia)* (1997) ICJ Reports 7, 78 (para. 140).
[56]   *Responsibilities and Obligations of States Sponsoring Persons and Entities with Respect to Activities in the Area* (n 14) para. 117.
[57]   Ibid.
[58]   Ibid., paras 117–120.
[59]   Rio Declaration (n 23) Principle 11.

establishing quantitative limits on atmospheric emissions of waste gases such as sulphur and nitrogen oxides (SOx and NOx),[60] chlorofluorocarbons (CFCs),[61] and carbon dioxide (CO_2),[62] few binding international obligations exist calling for limits on the generation of municipal and industrial waste.[63]

Nevertheless, underlying the Basel regime is the express recognition of the need to protect human health and prevent environmental harm through the reduction and minimization of hazardous wastes.[64] Reaffirmed in Agenda 21[65] and the 2002 Plan of Implementation of the World Summit on Sustainable Development (WSSD),[66] the concept of waste minimization lies at the heart of the contemporary movement to 'Reduce, Reuse, Recycle'. The Basel Convention positively obliges States to ensure that the generation of hazardous and other wastes is reduced to a minimum taking into account social, technological and economic impacts,[67] and to prevent or minimize the consequences of pollution arising from the management of hazardous or other wastes.[68] Although light on specific details as to how to achieve waste minimization, in requiring parties to keep their wastes at home, the proximity principle, which requires waste to be managed and disposed of as close as possible to the point of generation,[69] is intended to operate to drive up the cost of waste disposal thereby producing economic incentives for pollution prevention and reduced waste generation.[70] This operation of the proximity principle as a manifestation of the preventive principle is evident, for example, in the 2002 Strategic Plan for the Implementation of the Basel Convention, which called for the 'active promotion and use of cleaner technologies and production, with the aim of the prevention and minimization of hazardous and other wastes subject to the Basel Convention'.[71]

---

[60] Convention on Long Range Transboundary Air Pollution (n 21).

[61] Montreal Protocol on Substances that Deplete the Ozone Layer (n 21).

[62] Kyoto Protocol (n 21).

[63] Sands and Peel (n 54) 560.

[64] Basel Convention (n 20) Preamble.

[65] Agenda 21, Report of the United Nations Conference on Environment and Development, I (1992) UN Doc A/CONF.151/26/Rev.1, Chs 20 and 21.

[66] The WSSD Johannesburg Plan of Implementation is available at <www.un.org/esa/sustdev/documents/WSSD_POI_PD/> (last accessed on 13 August 2015).

[67] Basel Convention (n 20) Art. 4 (2)(a).

[68] Ibid., Art. 4 (2)(a) and (c).

[69] Ibid., Art. 4 (2)(b).

[70] David Hunter, James Salzman and Durwood Zaelke, *International Environmental Law and Policy* (Foundation Press, 4th ed. 2011) 953.

[71] See <www.basel.int/stratplan/index> (last accessed on 13 August 2015).

In addition to the principle of minimization of waste, prevention is further evident in the requirement that wastes be managed and disposed of in an environmentally sound manner. Defined in the Basel Convention as meaning 'taking all practicable steps to ensure that hazardous wastes or other wastes are managed in a manner which will protect human health and the environment against the adverse effects which may result from such wastes',[72] the principle of environmentally sound management applies to waste disposal both within the jurisdiction of the generating State and in importing States. With respect to the former, parties are to ensure the availability of adequate disposal facilities for the environmentally sound management of hazardous and other wastes which, by operation of the proximity principle, are to be located as close as possible to the source of the waste.[73] With respect to the latter, exporting parties must require that wastes to be exported are managed in an environmentally sound manner in the State of import and any transit States,[74] while potential importing parties must prevent imports where they have reason to believe they will not be managed in an environmentally sound manner.[75] Under no circumstances can a party transfer its obligation to carry out environmentally sound management to other States although, per contra, it may impose additional requirements, consistent with the Convention, to better protect human health and the environment.[76]

Beyond the requirements of environmentally sound management, the Basel Convention provides further guidance on the content of due diligence by requiring, for example, that transport and disposal of hazardous and other wastes may only be carried out by authorized persons and that transboundary movements must conform with generally accepted and recognized international rules and standards of packaging, labelling and transport, and take account of relevant internationally recognized practices. Transboundary movements must also be accompanied by a movement document from the point of exit to the point of disposal.[77] Illegal traffic of hazardous or other wastes must be considered a criminal activity and appropriate legal, administrative and other measures must be adopted to implement the provisions of the Convention and to prevent and punish its contravention.[78] Given the temporal nature of the

---

[72]   Basel Convention (n 20) Art. 2 (8).
[73]   Ibid., Art. 4 (2)(b).
[74]   Ibid., Art. 4 (8).
[75]   Ibid., Art. 4 (2)(g).
[76]   Ibid., Art. 4 (10) and 4 (11).
[77]   Ibid., Art. 4 (6).
[78]   Ibid., Art. 4 (3) and 4 (4).

obligation of due diligence, the specific content of the obligations of waste minimization and environmentally sound management and the measures needed to ensure their achievement will vary over time as new threats to human health and the environment are identified and new approaches to waste management, such as the integrated life-style approach,[79] are developed.

1.5   THE PRINCIPLE OF COOPERATION

The obligation on States to cooperate in addressing international issues is recognized as a fundamental rule of general international law emanating from the principle of 'good-neighbourliness' enunciated in Article 74 of the UN Charter.[80] Principle 24 of the Stockholm Declaration and Principle 27 of the Rio Declaration confirm the obligation on States to cooperate 'in good faith and in a spirit of partnership' in all matters concerning protection of the environment. While the precise nature and extent of the obligation remains a matter of contestation,[81] its customary status, at least, is not contested.[82] However, it is important to remember that the obligation to cooperate does not mandate a specific outcome or the prior consent of potentially affected states.[83] Principle 14 of the Rio Declaration merely requires States to cooperate 'effectively' to 'discourage or prevent the relocation and transfer to other states of any activities and substances that cause severe environmental degradation or are found to be harmful to human health', while Principle 19 merely requires States to 'provide prior and timely notification and relevant information to potentially affected states on activities that may have a significant transboundary environmental effect and to consult with those states at an early stage and in good faith'. Rather, as Principle 19 indicates, the proper observance of the principle of cooperation (merely) requires fulfilment of certain procedural obligations such as those relating to environmental assessment, exchange of information, notification, consultation

---

[79]   As called for in Agenda 21 (n 65) Ch 20, paras 20.1, 20.2 and 20.6.
[80]   Charter of the United Nations (adopted on 26 June 1945, entered into force on 24 October 1945) 1 UNTS xvi.
[81]   *Pulp Mills* (n 13).
[82]   See, e.g., *Gabčikovo-Nagymaros* (n 55) paras 141–142; *Mox Plant (Ireland v UK)* (Provisional Measures) ITLOS, Order of 3 December 1981, para. 83.
[83]   *Lac Lanoux* Arbitration *(France v Spain)* (1957) 12 RIAA 281; 24 ILR 101 and *Pulp Mills* (n 13).

and negotiation 'on the basis of the principle of good faith and in the spirit of good neighbourliness'.[84]

The requirements of cooperation are manifest in the Basel Convention in its provisions relating to, for example: notification to the Secretariat of national definitions of hazardous wastes; notification to other parties of decisions to prohibit imports; information exchange on transboundary movements and the potential and actual effects thereof on human health and the environment; dissemination of information on transboundary movements for the purpose of improving environmentally sound management and preventing illegal traffic; and information exchange on technical and scientific know-how, on sources of advice and expertise, and on the availability and capabilities of sites for disposal to States concerned.[85] However, the Basel Convention mandates a wholly new mode of cooperation, far more stringent than the mere consultation and notification requirements generally required by the principle of cooperation.

Embodied in the principle of 'prior informed consent', the Convention mandates the explicit prior consent of potentially affected States, a consent that must be based on information supplied by an exporter, which must be sufficient to enable the nature and the effects on human health and the environment of the proposed movement to be assessed. The importing State is then at liberty either to consent to the shipment, with or without conditions, or deny permission, or request additional information pending a final decision. In the absence of such consent and an agreement between the exporting State and the disposer specifying environmentally sound management of the waste in question, the State of export must not allow the transboundary movement to proceed. Transit States can also prohibit transit passage and export must not proceed unless and until their consent is obtained. Where consent is not obtained, or a transboundary movement cannot be completed, the exporting State is required to take back the waste unless alternative arrangements cannot be made for its environmentally sound management.[86] Any movement that takes place in violation of these requirements is to be considered illegal traffic and punished as a

---

[84]   As in the language of Principle 7 of the 1978 UNEP Draft Principles of Conduct for the Guidance of States in the Conservation and Harmonious Exploitation of Natural Resources Shared by Two or More States, available at <www.unep.org/Documents.multilingual/Default.asp?DocumentID=65&ArticleID=1260&l=en> (last accessed on 13 August 2015).

[85]   Basel Convention (n 20) Arts 3, 4 (1)(a), (f), (h).

[86]   Ibid., Art. 6.

criminal offence.[87] Similar provisions on prior informed consent are also found in the Bamako and Waigani Conventions.

This invocation of the principle of prior informed consent in the Basel Convention, and in other conventions dealing with trade in toxic or hazardous substances or wastes,[88] constitutes a far-reaching restriction on their trade and can be taken as powerful evidence of the recognition, in international law, of the shared responsibility of importing and exporting States for the protection of human health and the environment. Given that the principle is essentially an expression of State sovereignty, its customary status, at least in the context of the transboundary movement and disposal of toxic or hazardous wastes, seems accepted.[89]

1.6 THE PRINCIPLE OF SUSTAINABLE DEVELOPMENT

The general principle that States should ensure the development and use of their resources in a manner that is sustainable has been known in international law since at least the 1893 *Bering Sea Fur Seals* arbitration.[90] However, the specific term 'sustainable development' finds it origins in the 1987 Bruntland Report.[91] Defined there as meaning 'development that meets the needs of the present generation without compromising the ability of future generations to meet their own needs', sustainable development is perhaps best understood not as a specific principle of international law but rather as the end goal or final objective of human activities,[92] a goal which is to be pursued through the implementation of the various distinct legal principles embodied, for example, in the Rio Declaration and the 2002 WSSD Plan of Implementation. The ICJ refers to the term as a 'concept' rather than a principle, [93] and debate continues as to its

---

[87]   Ibid., Art. 9.

[88]   See, e.g., the Rotterdam Convention on Prior Informed Consent Procedure for Certain Hazardous Chemicals and Pesticides in International Trade (adopted on 10 September 1998, entered into force on 24 February 2004) (1999) 38 ILM 1.

[89]   Birnie, Boyle and Redgwell (n 6) 476–7 and 486.

[90]   *(Great Britain v United States)* (1893) 1 Moore's International Arbitration Awards 755.

[91]   Report of the World Commission on Environment and Development, *Our Common Future* (1987) 43.

[92]   See, e.g., Alan Boyle and David Freestone, 'Introduction' in Alan Boyle and David Freestone (eds), *International Law and Sustainable Development: Past Achievements and Future Challenges* (Oxford University Press 1999) 1.

[93]   *Gabčikovo-Nagymaros* (n 55) para. 140.

normativity. Thus, while the objective of sustainable development may be to reconcile economic development with protection of the environment, the extent to which the concept can legally constrain the behaviour of States is debatable.

Nevertheless, this does not mean that the concept lacks any legal function. In the *Gabčíkovo-Nagymaros* case the ICJ held that new norms and standards, including the concept of sustainable development, had to be taken into consideration and given proper weight both when contemplating new activities and when continuing activities begun in the past.[94] In other words, sustainable development can be considered a factor orienting the behaviour of States and guiding the interpretation of relevant rules in the judicial process.[95] In this respect, it reflects a range of procedural and substantive commitments and obligations, most notably those relating to the sustainable use of natural resources, intergenerational equity, and integration of environmental considerations into economic and other development.

As the name implies, the principle of sustainable use recognizes that limits on the rate of use or manner of exploitation of natural resources are necessary to ensure attainment of both the intra and intergenerational objectives of sustainable development. What those limits might be is a matter for determination by States acting cooperatively. In the waste context, this is reflected, in particular, in the recognition of the need for waste minimization and the prevention or minimization of the consequences of pollution arising from the management of hazardous or other wastes. The exemption from the Basel regime of wastes destined for recycling or recovery is further evidence of the desire of the parties to ensure sustainable use of their resources, a desire that was made manifest in the Cartagena Declaration on the Prevention, Minimization and Recovery of Hazardous Wastes and Other Wastes adopted by the Conference of the Parties in 2011.[96]

The point of sustainable use is not only to preserve resources for current, but also for future generations. Indeed, intergenerational equity is a fundamental aspect of the concept of sustainable development. However, intergenerational equity is not merely about preserving resources for future use but also implies the need to pass on to future generations a clean

---

[94]   Ibid.

[95]   Vaughan Lowe, 'Sustainable Development and Unsustainable Arguments' in Boyle and Freestone (eds) (fn 92) 19.

[96]   Report of the Conference of the Parties to the Basel Convention on the Control of Transboundary Movements of Hazardous Wastes and their Disposal on its tenth meeting, Doc UNEP/CHW.10/28, 1 November 2011, Annex IV.

and healthy environment. As Principle 1 of the Stockholm Declaration puts it, humans bear 'a solemn responsibility to protect and improve the environment for present and future generations'. Even while associating intergenerational equity with the right to development, Principle 4 of the Rio Declaration requires that right to be fulfilled 'so as to equitably meet developmental and environmental needs of present and future generations'. The elimination of 'toxic colonialism' through the export of environmental problems is, as noted at the outset, the fundamental *raison d'être* of the international legal regime for the transboundary movement of hazardous waste. When coupled with the requirements of self-sufficiency, proximity and the environmentally sound management of wastes by both generating and importing States, the regime provides strong environmental safeguards for both current and future generations.

The principle of integration, articulated in Principle 4 of the Rio Declaration and confirmed in the *Iron Rhine* case as a requirement of international law,[97] requires the integration of appropriate environmental measures into the design of economic development activities. As applied by the ICJ in the *Gabčikovo-Nagymaros* case, implementation of the principle requires the collection and dissemination of environmental information and the conduct of environmental impact assessments. These elements are reflected in the Basel regime in the many obligations on parties to collect and disseminate, either unilaterally or through the Secretariat, information on the hazardous (or otherwise) nature of wastes and to cooperate in the dissemination of information regarding transboundary movements and the monitoring of effects on human health and the environment, as well as any accidents which are likely to present risks to human health or the environment.[98] While not explicitly stated, the requirement of at least some form of environmental impact assessment is implicit in the requirement that notifications regarding potential transboundary movements include sufficient information to enable the nature and the effects on health and the environment of the proposed movement to be assessed.[99]

In some ways the principle of integration lies at the heart of the concept of sustainable development, which has always been articulated in terms of requiring States to ensure their development is compatible with the need to protect and improve the environment. In this respect, it is the integration principle which is said to serve as a basis for requiring 'green

---

[97] See in the *Iron Rhine* case (n 51) paras 59 and 243.
[98] I.e., Basel Convention (n 20) Arts 3, 10, 13.
[99] Ibid., Art. 6.

conditionality' in development assistance agreements.[100] Importantly for this volume, the integration principle also serves as a basis for the concept of the Green Economy and its support for the environmentally sound recycling and reclamation of valuable materials that can 'provide both economic opportunities and substantial environmental benefits by reducing the need to exploit non-renewable natural resources that might otherwise be mined in the absence of recycled materials'.[101]

1.7   THE PRECAUTIONARY PRINCIPLE

The final general principle considered here is the precautionary principle. In the international context, the precautionary principle – or approach, as it is also referred to – is of relatively recent vintage. The core of the principle is articulated in Principle 15 of the Rio Declaration, which states that 'where there are threats of serious or irreversible damage, lack of full scientific certainty shall not be used as a reason for postponing cost-effective measures to prevent environmental degradation'.[102] Importantly, Principle 15 also states that 'the precautionary approach shall be widely applied by states according to their capabilities'.

Despite its adoption in numerous environmental treaties and its invocation in international judicial and arbitral proceedings, neither the meaning nor the effect of the precautionary principle is yet agreed. On the one hand, it is argued that the principle provides the basis for early action to address threatening environmental issues. On the other hand, it is argued that application of the principle results in over-regulation and unwarranted limitations on human activity. Conflicting interpretations of the principle range from the requirement merely to act carefully when taking decisions that may have an adverse impact on the environment, to the requirement to regulate and possibly even prohibit activities and substances which may be environmentally harmful even in the absence of conclusive proof of such likely harm, to the requirement that the person wishing to carry out a particular activity must prove it will not cause environmental harm.[103] This latter interpretation, in particular, requires polluters to establish that their activities will not adversely affect the environment before they can

---

[100]   Sands and Peel (n 54) 667.
[101]   Hunter, Salzman and Zaelke (n 70) 943.
[102]   For comprehensive examinations of the precautionary principle in international law see e.g. David Freestone, *The Precautionary Principle: The Challenge of Implementation* (Kluwer Law International 1996) and Trouwborst (n 12).
[103]   Sands and Peel (n 54) 220.

be authorized to undertake the proposed activity, thus raising the connection between precaution and the requirements of environmental impact assessment.

Given these interpretive quandaries, it is perhaps not surprising that the status of the precautionary principle as a rule of customary international law remains uncertain. In the *Pulp Mills* case, the ICJ declined to comment on its customary status, stating only that 'a precautionary approach may be relevant in the interpretation and application of' the relevant treaty.[104] More recently, the ITLOS Seabed Disputes Chamber has held that the precautionary principle is 'an integral part of the general obligation of due diligence'[105] and that its incorporation into numerous treaties and other instruments has 'initiated a trend towards making this approach part of customary international law'.[106]

Regardless of the lack of certainty as to the meaning, effect and customary status of the precautionary principle, it is clear that the Basel Convention reflects 'a strong form of the precautionary approach'[107] by allowing States to refuse to accept waste and by requiring a State of export to demonstrate that the wastes will be managed in an environmentally sound manner before any export can go ahead. The burden is shifted to the proponent of the activity to satisfy not only importing States but also any transit States that the proposed waste movement will not cause environmental harm. The same approach is also evident in the Bamako and Waigani Conventions.

1.8   CONCLUSION

In the absence of principles, international law is, at best, a set of arbitrary rules; at worst, a theoretical hoax of international lawyers. The general principles of international environmental law discussed in this chapter provide the critical theoretical foundation for the normative framework that has developed in international law regarding the management and transboundary movement of waste. However, as noted at the outset, the effectiveness of international law cannot be secured by principles alone. Rather, principles are only one element in an international system that requires recognition of the inter-linkages between principles, specific rules

---

104   *Pulp Mills* (n 13) para. 164.
105   *Responsibilities and Obligations of States* (n 14) para. 131.
106   Ibid., para. 135.
107   Birnie, Boyle and Redgwell (n 6) 473.

and institutional mechanisms for securing compliance. This interactive process is particularly evident in the international regime relating to the management and transboundary movement of waste where operationalization of the basic principles of permanent sovereignty, no harm, prevention, cooperation, sustainable development and precaution has led to the development of new principles, such as those of self-sufficiency, proximity, waste minimization, environmentally sound management and prior informed consent, all of which are further operationalized in the detailed rules set out in the Basel Convention and other treaties dealing with waste management. Thus, while insufficiently detailed in themselves to create binding legal obligations, these principles provide valuable interpretive guidance both as to the manner in which the law has been developed and applied and, thanks in particular to their interpretational flexibility, as to the manner in which the law should continue to be developed and applied into the future.

2. Waste and international law: towards a resource-based approach?

Tarcísio Hardman Reis[1]

EXECUTIVE SUMMARY

The present chapter provides an overview of the treatment of wastes in international law through a study of international and regional treaties, as well as some of the existing jurisprudence, in order to identify trends and gaps related to the international regulation of waste. Within this purpose, the chapter identifies three approaches based on different topics under international law: the protection of human rights; the protection of the environment; and economic concerns associated with trade and investment activities. The chapter allows us to observe that each of the approaches described serves to respond to specific concerns (e.g. the nuisances created by waste, pollution from certain types of waste, and technical and legal definitions). The chapter concludes that an economic approach, mainly supported by soft law instruments (e.g. international standards and publications from international organizations) is currently being developed in order to respond to the growing importance of the economic dimension of waste.

2.1 INTRODUCTION

The document 'The Future We Want'[2] notes that lower negative environmental impacts, increased resource efficiency and waste reduction are objectives of the Green Economy.[3] Among other things, it recognizes the importance of adopting a life-cycle approach, and of further development

---

[1] The views expressed herein are those of the author and do not necessarily reflect the views of the United Nations.

[2] United Nations, 'The Future We Want', UNGA Res. 66/288 (27 July 2012) A/RES/66/288.

[3] Ibid., para. 60.

and implementation of policies for resource efficiency and environmentally sound management. It also proposes commitments in relation to the reduction, reuse and recycling of wastes 'with a view to managing the majority of global waste in an environmentally sound manner and, where possible, as a resource'.[4]

Although at a first glance the declaration provides an overview of concepts that are not new to environmental negotiations,[5] it is interesting to note the inclusion of waste as a resource, albeit with noticeable hesitation. Linking waste and resources is a result of long-term efforts of the international community to define waste, and it may represent an important shift in the way wastes are regulated under international law and consequently transposed into domestic law.

The international community has been long and increasingly concerned with the toxic characteristics of certain types of substances,[6] which are marked by a vague, yet fundamental, distinction between hazardous and non-hazardous wastes. Despite a remarkable growth in waste generation, the importance given to hazardous wastes[7] was not followed by the development of an adequate international approach to non-hazardous wastes.

In fact, the economic implications of waste are often disregarded and its legal implications overlooked when, for example, recoverable materials may qualify as tradable commodities despite posing considerable risk to human health and the environment.[8] International governing bodies apply regulations inconsistently to wastes, requiring certain types to be subject to strict international regulations, while others benefit

---

[4] Ibid., para. 218.

[5] E.g. Bali Declaration on Waste Management for Human Health and Livelihood, Conference of the Parties to the Basel Convention on the Control of Transboundary Movements of Hazardous Wastes and their Disposal, Ninth Meeting (Bali, 23–27 June 2008), UN Doc UNEP/CHW.9/39, paras 4 and 5; and the Cartagena Declaration on the Prevention, Minimization and Recovery of Hazardous Wastes and Other Wastes, Conference of the Parties to the Basel Convention on the Control of Transboundary Movements of Hazardous Wastes and their Disposal, Tenth Meeting (Cartagena, Colombia, 17–21 October 2011) UN Doc UNEP/CHW.10/28, paras 1, 2 and 6

[6] Louis B. Sohn, 'The Stockholm Declaration on the Human Environment' (1973) 14 *Harvard International Law Journal* 423. See comment on Principle 6 at 462.

[7] UNEP, Montevideo Programme for the Development and Periodic Review of Environmental Law, Decision 10/21 of the Governing Council (31 May 1982) 4.

[8] Marina Grosz, *Sustainable Waste Trade under WTO Law: Changes and Risks of the Legal Frameworks' Regulation of Transboundary Movements of Wastes* (Brill/Nijhoff 2011) 267.

from the trade liberalism mechanism promoted by the World Trade Organization.[9] This may be the underlying reason for the absence of a coherent international legal framework adequately covering, for example, waste-minimization strategies, production, and consumption.[10]

Nonetheless, wastes are governed by international law, including a number of international treaties, such as the Basel Convention,[11] and soft law mechanisms. Most importantly, international law also governs wastes in relation to its potential to cause harm to human beings and to interfere with human rights. This anthropocentric approach serves as the fundamental base for a diverse range of international obligations that deserve to be carefully analysed.

2.2 A RIGHTS-BASED APPROACH TO THE MANAGEMENT OF WASTES

Defining a straightforward link between waste and pollution is challenging. Many waste-related activities, such as the management and production of organic waste, present low risks to human health. Other activities, such as recycling and incineration, may also be conducted safely as long as preventive actions are adequately taken. Therefore, in the rights-based approach, the distinction between hazardous and non-hazardous wastes is made based on how they affect the enjoyment of human rights.

Theoretically, a rights-based approach to waste-related activities is formed with the application of specific recognized individual rights that generate obligations for states to take the necessary actions to protect those rights. In this regard, fundamental rights are an important source of legal protection to individuals from cases of severe pollution caused by waste.

It is convenient, therefore, to analyse waste through its potential interference of the full enjoyment of human rights as an initial step to this study. This approach will allow the analysis of relevant materials and jurisprudence specific to human rights. It will also help to identify the anthropocentric foundations of different international materials related directly or indirectly to waste. In fact, the right to life is a universally

---

9 Ibid., 509.
10 Ibid., 515.
11 Basel Convention on the Control of Transboundary Movements of Hazardous Wastes and Their Disposal (adopted on 22 March 1989, entered into force on 5 May 1992) 1673 UNTS 57.

recognized right,[12] formulated in different international texts,[13] and it may benefit individuals against the adverse effects of pollution (2.2.1). Additionally, certain circumstances may also generate specific obligations for states based on other fundamental rights (2.2.2).

2.2.1   The Right to Life as a Mechanism of Protection Against Certain Polluting Activities

The most basic element in a case of infringement of the enjoyment of the right to life is that an individual, or a group of individuals, are exposed to a type of pollution that is considered detrimental to health. This is possible either through a one-time event, for example with the pollution originating from an industrial accident of grave proportions, or through continuous exposure to toxic emissions.

Again, a distinction should be made between harmful and non-harmful activities to human life. By reaffirming that 'the illicit traffic and the dumping of toxic and dangerous products and wastes constitute a serious threat to the human right to life and health of every individual',[14] the Commission on Human Rights emphasized that specific types of activities, namely illegal traffic and the dumping of toxic wastes, require governments to take adequate legislative measures.[15] In another instance, the Inter-American Court of Human Rights considered that the environmental pollution produced by a field of toxic waste sludge next to the community of San Mateo, in Peru, was a violation of Article 4 of the American Convention on Human Rights.[16]

In addition to operations involving toxic waste, it is relevant to consider that a number of industrial activities may also negatively impact the enjoyment of the right to life. When it comes to extractive activities, for example, the United Nations Special Rapporteur on Toxic Waste highlighted that:

---

[12]   Universal Declaration of Human Rights, UNGA Resolution 217A (III) (1948) UN Doc A/810, Art. 3.

[13]   Convention for the Protection of Human Rights and Fundamental Freedoms (European Convention on Human Rights, as amended) (ECHR) Art. 2; American Declaration of the Rights and Duties of Man, OAS Res. XXX adopted by the Ninth International Conference of American States (1948) reprinted in Basic Documents Pertaining to Human Rights in the Inter-American System, OEA/Ser L V/II.82 Doc 6 Rev. 1 Art. 4.

[14]   CHR, Res. 1995/81 (51st Session, 8 March 1995) UN Doc E/CN.4/RES/1995/81, para. 2.

[15]   Ibid., para. 4.

[16]   Inter-American Court of Human Rights, *Community of San Mateo de Huanchor and its Members v. Peru*, Case 504/03, Report No. 69/04, Inter-Am. C.H.R., OEA/Ser.L/V/II.122 Doc 5 Rev. 1 at 487 (2004).

Extractive activities typically result in the introduction of hazardous substances in the natural environment, which may or may not be the desired resource, with impacts to human health, the environment, and society. The impacts of hazardous substances and waste on human life may occur through various paths of exposure, such as inhalation (. . .), ingestion (. . .), and physical contact with chemicals.[17]

As for the protection of the right of life, the approach of human rights is to 'lay down a positive obligation on States to take appropriate steps to safeguard the lives of those within their jurisdiction'.[18] The link between the role of the state and a harmful activity is therefore a key element to understanding the rights-based approach to wastes. For example, in the case *EHP v. Canada*[19] concerning the storage of radioactive waste near residential areas, the United Nations Human Rights Committee referred to the 'obligation of State parties to protect human life'.

Part of this obligation to protect human life was defined by the United Nations Rapporteur on Toxic Waste, 'The right to life involves at least a prohibition on the State not to take life intentionally or negligently. Thus, in extreme cases, the right can be invoked by individuals to obtain compensation where death results from some environmental disasters (. . .)'.[20]

Based on this definition it is convenient to seek further clarity on the distinction between intention and negligence of the state in the protection of human rights.

In the case *Öneryildiz v. Turkey*,[21] the European Court of Human Rights (ECtHR) examined a case where a Turkish national filed a complaint over the death of family members due to a methane explosion in a rubbish tip located near the illegally placed dwellings where he lived with his family. Upon review of the case, the Court decided that the Turkish authorities had known or ought to have known that there was a real or immediate risk to persons living near the rubbish tip, and that they had an obligation under Article 2 of the European Convention on Human Rights

---

[17] Human Rights Council, 'Report of the Special Rapporteur on the human rights obligations related to environmentally sound management and disposal of hazardous substances and waste, Calin Georgescu' (2012) UN Doc A/HRC/21/48, para. 20.

[18] *Öneryildiz v. Turkey*, 39 EHRR 12 para. 71 (2004).

[19] *EHP v. Canada*, Communication No. 67/1980, UN Doc CCPR/C/OP/1 (1982).

[20] Human Rights Council, 'Adverse effects of the illicit movement and dumping of toxic and dangerous products and wastes on the enjoyment of human rights – Report of the Special Rapporteur, Okechukwu Ibeanu' (20 February 2006) UN Doc E/CN.4/2006/42, para. 36.

[21] *Öneryildiz v. Turkey* (n 18) paras 71–73.

(ECHR) to take such preventive operational measures as were necessary and sufficient to protect those individuals.

Based on the recognition of the right to life as a fundamental right in different international instruments, states are responsible for making use of their regulatory power and administrative ability to exercise reasonable control of respective industries.[22]

2.2.2 Other Existing Human Rights that May be Affected by Polluting Activities

The protection of the right to life is not the only right opposable to states when it comes to polluting activities. For example, the improper management and disposal of medical waste may pose a significant threat to the right to the highest attainable standard of physical and mental health, the right to safe and healthy working conditions, and the right to an adequate standard of living, in addition to the right to life.[23] Another example is mining wastes, which may be detrimental to the right to adequate food and nutrition; the right to a safe and healthy working environment; the right to safe drinking water and adequate sanitation; and the right to enjoyment of a safe, clean and healthy sustainable environment.[24] In addition, the illicit movement of toxic and dangerous products may also be subject to the role of the state in protecting other individual rights such as the rights to food, adequate housing, clean water, and safe and healthy working conditions.[25]

Without providing a comprehensive assessment of all the possible types of rights that may affect different waste-related activities, it is opportune to limit the present analysis to the applicable case law developed by the ECtHR in relation to the right to respect for private and family life, as well as to the relevance of the right to information and participation in relation to waste-related activities.

In relation to the right to respect for private and family life, the ECHR states in its Article 8 that '[e]veryone has the right to respect for his private and family life, his home and his correspondence'. Paragraph 2 of the

---

[22]   Dimitris Xenos, 'Asserting the Right to Life (Article 2, ECHR) in the Context of Industry' 8 *German Law Journal* 231, 252.

[23]   Human Rights Council, 'Report of the Special Rapporteur on the adverse effects of the movement and dumping of toxic and dangerous products and wastes on the enjoyment of human rights, Calin Georgescu' (4 July 2011) UN Doc A/HRC/18/31, para. 18.

[24]   Human Rights Council, 'Report 2012' (n 17).

[25]   Human Rights Council, 'Report 2006' (n 20) para. 17.

same article reads, '[t]here shall be no interference by a public authority with the exercise of this right except such as is in accordance with the law and is necessary in a democratic society in the interests of national security, public safety or the economic well-being of the country (. . .)'.

In the case *Lopez Ostra v. Spain*,[26] the ECtHR held that a failure of the state to protect the home, private and family life of one of its citizens from the pollution caused by a waste treatment facility was a violation of Article 8, where there was a sufficiently serious interference with the applicants' enjoyment of their home and private life. The Court recognized a fair balance between the competing interests of the individual and of the community as a whole, which gives states a certain margin of appreciation.[27]

In a subsequent case[28] related to a complaint from the local population that the authorities had not taken appropriate action to reduce the risk of pollution from a chemical factory, the Court clarified that there may be positive obligations inherent in effective respect for private and family life. It concluded that the respondent state had not taken the necessary steps to ensure effective protection of the applicant's right to respect for their private and family life.[29]

In a later case,[30] upon the review of a case related to the pollution resulting from a plant for the storage and treatment of 'special waste' classified as either hazardous or non-hazardous, the ECtHR reaffirmed the margin of appreciation of states in striking a fair balance between the competing interests of the individual and of the community. The Court recognized that it is for the national authorities to make the initial assessment of the necessity for interference. It noted that states are, in principle, better placed than an international court to assess the requirements relating to the treatment of industrial waste in a particular local context and to determine the most appropriate environmental policies and individual measures while taking into account the needs of the local community.[31]

Some lessons from the European case law in relation to the right to respect for private and family life are the criteria established for the margin of appreciation by the states in reaching a fair balance between individual and collective rights, and the role of states in making an initial assessment of the necessity of interference in relation to the treatment of industrial wastes.

---

[26] *Lopez Ostra v. Spain* (1994) 20 EHRR 277.
[27] Ibid., para. 51.
[28] *Guerra v. Italy* (1998) 26 EHRR 357.
[29] Ibid., para. 58.
[30] *Giacomelli v. Italy* (2007) 45 EHRR 38.
[31] Ibid.

A similar balance is also necessary in relation to the right to information and participation. It was noted that many disputes related to the movement of toxic and dangerous wastes arise due to a lack of information and the failure of the states or corporations to disclose the potential dangers of certain activities.[32] The United Nations Special Rapporteur on Toxic Wastes underlined that states may only invoke grounds of national security, trade secrets and confidentiality insofar as they are in conformity with relevant derogation or limitation clauses of international human rights instruments.[33]

It is also noted that the Aarhus Convention on Access to Information, Public Participation in Decision-making and Access to Justice in Environmental Matters[34] adopts a rights-based approach to the issue of access to information by requiring parties to guarantee access to information in environmental matters. It refers to the goal of protecting the right of every person of present and future generations to live in an environment adequate to health and well-being. The Aarhus Convention also guarantees the right to public participation in decision-making processes relating to environmental matters, which is essential to secure a rights-based approach to the regulation of toxic chemicals.[35]

Noting that the right to respect for private and family life is not universally guaranteed, and that the right of information and participation faces many obstacles to its general implementation, it is possible to assert that individual rights are the basis for states to impose limits on waste-related activities, whether these activities are undertaken by the states themselves or through private agents.

In sum, a rights-based approach has been developed in order to prevent waste-related operations interfering with the enjoyment of human rights. By establishing a balance between the interests of individuals and of a state, this approach does not ignore the economic aspects of waste and in particular the potential socio-economic benefits from waste-related operations.

---

[32] Human Rights Council, 'Report of the Special Rapporteur on the adverse effects of the illicit movement and dumping of toxic and dangerous products and wastes on the enjoyment of human rights, Okechukwu Ibeanu' (18 February 2008) UN Doc A/HRC/7/21, para. 34.

[33] Ibid., para. 36.

[34] Convention on Access to Information, Public Participation in Decision-Making and Access to Justice in Environmental Matters (Aarhus Convention) (adopted on 25 June 1998, entered into force on 30 October 2001) 2161 UNTS 447.

[35] Human Rights Council, 'Report 2006' (n 20) para. 41.

2.3 A TREATY-BASED APPROACH TO THE MANAGEMENT OF WASTES

As well as respecting fundamental rights, states are also required to respect their international commitments pertaining to the environment. Many of the international obligations covering waste and certain related activities are defined in environmental agreements, notably the Basel Convention.[36] A diverse range of international treaties with a focus on issues such as the protection of the sea;[37] the international regulation of radioactive activities;[38] and the protection of workers[39] also contain relevant provisions on wastes. In addition, a number of related regional instruments are in force,[40] including within the European Union (EU).[41] As the content of these instruments is very diverse, a comparative approach helps to analyse some of the commonalities and, in particular, highlight the trade-control provisions (2.3.1) and the waste definitions (2.3.2) contained in the existing treaties.

---

[36] Basel Convention (n 11).

[37] International Convention for the Prevention of Pollution from Ships (MARPOL Convention), as amended (1978) (adopted on 17 February 1978, entered into force on 2 October 1983) 1340 UNTS 61.

[38] Joint Convention on the Safety of Spent Fuel Management and on the Safety of Radioactive Waste Management (adopted on 24 December 1997, entered into force on 18 June 2001) 36 ILM 1431.

[39] Convention concerning the Protection of Workers against Occupational Hazards in the Working Environment due to Air Pollution, Noise and Vibration (ILO No. 148) (adopted on 20 June 1977, entered into force on 11 July 1979) 1141 UNTS 108; Convention concerning Safety in the Use of Asbestos (ILO No. 162) (adopted on 24 June 1986, entered into force on 16 June 1989) 2 SMT 359; Convention on Safety in the Use of Chemicals at Work (ILO No. 170) (adopted on 24 June 1990, not in force) 1753 UNTS 189.

[40] Bamako Convention on the Ban of the Import into Africa and the Control of Transboundary Movement and Management of Hazardous Wastes within Africa (Bamako Convention) (adopted on 30 January 1991, entered into force on 22 April 1998) 30 ILM 773; Convention to Ban the Importation into Forum Island Countries of Hazardous and Radioactive Wastes and to Control the Transboundary Movement and Management of Hazardous Wastes within the South Pacific Region (Waigani Convention) (adopted on 16 September 1995, entered into force on 21 October 2001) 2161 UNTS 91; Convention for the Protection of the Mediterranean Sea against Pollution (Barcelona Convention) (adopted on 16 February 1976, entered into force on 12 February 1978) 1102 UNTS 44, and amendments.

[41] EU Directive, regulation and decisions.

2.3.1   Control of the International Movement of Wastes

A number of instruments contain trade-related provisions for the control of the transboundary movement of wastes. A key aspect of this type of control is the prior informed consent mechanism, which is referred to in the Basel Convention as the notification procedure. This mechanism is meant to ensure that the competent authorities of the state of import and transit know that hazardous wastes are to be shipped, and to provide them with the details of that shipment. This notification system allows importing countries to authorize or prohibit the transboundary movement.[42] This type of mechanism is the cornerstone of the international governance on wastes. It is also present in the related regional instruments[43] and referred to in the Stockholm Convention.[44] It is worth noting that a similar mechanism exists in relation to the transboundary movement of radioactive waste.[45]

Other instruments propose alternative versions of this type of mechanism. For example, the OECD promotes a tracking system that divides wastes by their risks for human health and the environment during its transboundary movement.[46] The EU also adopts more detailed trade related provisions for shipments of hazardous wastes between its Member States.[47]

Additionally, other conventions contain provisions related to accidents during the transboundary movement. For example, the Waigani and the Bamako Conventions incorporate provisions related to the transmission of information in case of accidents during the transboundary movement[48] and the Barcelona Convention requires parties to 'take all appropriate measures to prevent, abate and to the fullest possible extent eliminate pollution of the environment which can be caused by transboundary movements and disposal of hazardous wastes, and to reduce to a minimum, and if possible eliminate, such transboundary movements'.[49]

---

[42]   Basel Convention (n 11) Art. 6.

[43]   Waigani Convention (n 40) Art. 6; Bamako Convention (n 40) Art. 6.

[44]   Stockholm Convention on Persistent Organic Pollutants (POPS) (adopted on 23 May 2001, entered into force on 17 May 2004) 2256 UNTS 119, Art. 6 (1) (d)(iv).

[45]   Michel Montjoie, *Droit international et gestion des déchets radioactifs* (LGDJ 2011).

[46]   OECD, Decision of the Council concerning the Control of Transboundary Movements of Wastes Destined for Recovery Operations (14 June 2001) C(2001)107/FINAL, and amendments.

[47]   Regulation (EC) 1013/2006 of 14 June 2006 on shipments of waste (2006) OJ L190/1.

[48]   Waigani Convention (n 40) Art. 7; Bamako Convention (n 40) Art. 13.

[49]   Barcelona Convention (n 40) Art. 11.

A different type of trade control is the prohibition of the trade. One example is the ban adopted by the Parties of the Basel Convention, prohibiting the transboundary movements of waste between OECD countries to non-OECD countries.[50] More strict prohibitions are made in the Bamako Convention, which 'prohibits the import of all hazardous wastes, for any reason, into Africa from non-Contracting Parties',[51] and in the Waigani Convention, which bans the import and export of all hazardous wastes and radioactive wastes from outside the area of the Convention.[52] As a response to this, the EU prohibits the export of different types of wastes to certain countries.[53]

This general prohibition is complemented by specific provisions set up, for example, by the Minamata Convention, which requires parties to the Basel Convention not to transport mercury waste across international boundaries, except for the purpose of environmentally sound disposal.[54] Without a specific reference to the Basel Convention, Article 3 of the Stockholm Convention goes in a similar direction by requiring that chemicals targeted for phase out (which in practical terms is tantamount to defining such chemicals as waste) be imported or exported only in very few circumstances, including for the purpose of environmentally sound disposal.[55]

An effective mechanism of trade control would not be possible without the clarification of the consequences related to the violation of trade restrictions. The Basel Convention and the Bamako Convention list different cases of illegal traffic, including the transboundary movement without notification; without the consent of the states concerned or obtained through falsification, misrepresentation or fraud; in nonconformity with the accompanying documents; or resulting in deliberate disposal.[56] The Waigani Convention adds a violation of the ban in the region of the Convention to this list.[57] The EU also included additional aspects in its definition of the illegal shipment.[58] As for its consequences, the Basel, the

---

[50] UNEP, Ban Amendment, Decision II/12, Second Meeting of the Conference of the Parties to the Basel Convention on the Control of Transboundary Movements of Hazardous Wastes and their Disposal (Geneva, 21–25 March 1994) UN Doc UNEP/CHW.2/30 (not in force).

[51] Bamako Convention (n 40) Art. 4.

[52] Waigani Convention (n 40) Art. 4.

[53] Regulation (EC) 1013/2006 (n 47) Arts 36, 39 and 40.

[54] Minamata Convention on Mercury (adopted on 10 October 2013, not in force) UN Doc UNEP(DTIE)/Hg/CONF/4, Art. 11(3)(c).

[55] Stockholm Convention (n 44) Art. 3 (2).

[56] Basel Convention (n 11) Art. 9; Bamako Convention (n 40) Art. 9.

[57] Waigani Convention (n 40) Art. 9.

[58] Regulation (EC) 1013/2006 (n 47) Art. 2.

Bamako and the Waigani Conventions establish alternative obligations for the state of export in cases of illegal traffic (i.e. a take-back procedure or disposal in accordance with the Convention).[59] The EU adopts a similar approach.[60]

An approach adopted by different conventions requires parties to introduce appropriate legislation to prevent and punish illegal traffic.[61] As a result, states may characterize additional types of activities and incorporate different consequences, including penalties and administrative sanctions, in cases of illegal traffic of hazardous wastes.

In sum, the control of the transboundary movement of wastes is a response to the potential adverse effects caused by certain types of wastes once their adequate management is not ensured. In this regard, international instruments provide for efficient control mechanisms that are applicable to the movement of certain types of waste.

On the other hand, it is important to note that non-hazardous wastes or wastes that are destined for recycling operations, such as metal scraps or chemicals that can be reutilized, are international commodities subject to free trade.[62] Consequently, there is a clear difference between the aim of existing agreements for the control of the transboundary movement of wastes and trade related agreements. As it was observed:

> while international trade law stipulates the importance of the unhindered transfer of wastes with the objective of promoting strong and specialized waste management industries with comparative advantages in terms of specialized know-how and technologies, as well as efficient treatment operations, international environmental law focuses particularly on the potentially polluting effects which such transboundary movements may have, and will tend to restrain cross-border waste trading.[63]

The differentiation between controlled waste and commodity is not based on value. In fact, some categories of controlled wastes can be very

---

[59]   Basel Convention (n 11) Art. 9; Bamako Convention (n 40) Art. 9; Waigani Convention (n 40) Art. 9.

[60]   Regulation (EC) 1013/2006 (n 47) Art. 24.

[61]   Basel Convention (n 11) Art. 9 (5); Bamako Convention (n 40) Art. 9 (2); Waigani Convention (n 40) Art. 9 (2).

[62]   In the system of the World Trade Organization, for example, the Technical Barriers to Trade (TBT) Agreement (15 April 1994) 18 ILM 1979, and the Agreement on the Application of Sanitary and Phytosanitary Measures (SPS Agreement) (15 April 1994) 33 ILM 1125, seek to ensure that requirements that products must fulfill for environmental purposes do not create unnecessary obstacles to international trade.

[63]   Grosz (n 8) 509.

valuable, such as mercury or copper scrap. It is also not exclusively related to its potential hazardous characteristics, as for example some traded goods like medicaments and equipment containing radioactive materials can be very harmful when not properly managed. The key to differentiating between the types of wastes that are subject to international controls, and commodities that are subject to free trade, is based on a system of definitions of wastes and non-wastes.

2.3.2 Waste-related Concepts and Definitions

One of the main obstacles towards considering waste to be a resource is the definition of waste in international law. First, none of the current regulations applicable to transboundary movements of waste define 'waste' in an abstract or exhaustive manner.[64] An adequate definition is important to distinguish wastes from non-waste, separate non-hazardous from hazardous wastes, and define the scope of legal instruments.

Beyond clarifying the scope of environmental instruments, an unambiguous differentiation between hazardous and non-hazardous wastes would provide the first step towards establishing efficient policies on trade and management of wastes,[65] which would take into account both economic and environmental aspects. In this context, a prolific jurisprudence has been developed through the European Court of Justice as to the definition of wastes and non-wastes.[66] In particular, the *Palin Granit* decision served to differentiate waste from by-products and wastes for re-use where a financial incentive is identified.[67] As shown by the evolution of the European jurisprudence, an ambiguous definition will have a negative impact by creating legal opacity and economic inefficiencies in a number of sectors concerned with waste production, recycling, and reutilization of resources.

When it comes to the differentiation between hazardous and non-hazardous wastes, the international definitions seem to be less ambiguous, primarily due to the work of the OECD and the evolution of international treaties, such as the Basel Convention. However, we will see that the definitions of hazardous wastes in international instruments may be confusing and even different if we compare, for example, the definitions of the Basel Convention with the definition of the International Convention on Liability and Compensation for Damage in Connection

---

[64] Ibid., 510.
[65] Ibid., 513.
[66] Martha Grekos, 'Finding a Workable Definition of Waste: Is It a Waste of Time?' (2006) *Journal of Politics and Law* 463.
[67] ECJ, *Case C-9/00 Palin Granit Oy* [2002] ECR I-3533.

with the Carriage of Hazardous and Noxious Substances by Sea (HNS Convention).[68]

Wastes are defined similarly in the Basel, Bamako and Waigani Conventions as 'substances or objects which are disposed of or are intended to be disposed of or are required to be disposed of by the provisions of national law'.[69] This approach was based on the OECD definition,[70] which evolved from a definition based on the waste trade into a comprehensive waste classification scheme.[71]

The OECD classification has been harmonized[72] with the Basel Convention, and with the Bamako and the Waigani Conventions[73] where hazardous wastes are defined as substances covered by the scope of each of these Conventions, which resulted in the direct association of the international definition of hazardous wastes with the scope of these Conventions, or at least the content of its annexes. For the Basel Convention, its scope is hazardous wastes belonging to 'any category contained in Annex I, unless they do not possess any of the characteristics

---

[68]   International Convention on Liability and Compensation for Damage in Connection with the Carriage of Hazardous and Noxious Substances by Sea (HNS Convention) (adopted 3 May 1996, not in force) 35 ILM 1406.

[69]   Basel Convention (n 11) Art. 2 (1).

[70]   OECD, Decision of the Council on Transfrontier Movements of Hazardous Wastes (27 May 1988) C(88)90/FINAL and amendment, Art. I (b)(ii).

[71]   By moving through the tables one by one, and selecting the code number in each table that best identified the purpose, destination, and characteristics of the waste in question, it is possible to arrive an International Waste Identification Code (IWIC) for every possible type of hazardous waste. By making all of this information readily accessible to carriers, receivers and customs officials, the IWIC is intended to facilitate the control of all hazardous wastes 'from generation to disposal', see Fred Morrison and William Muffet, 'Hazardous Waste' in Fred Morrison and Rüdiger Wolfrum (eds), *International, Regional and National Environmental Law* (Kluwer Law International 2000) 423.

[72]   OECD, Guidance Manual for the Implementation of the OECD Recommendation C(2004)100 on Environmentally Sound Management (ESM) of Waste (OECD 2007).

[73]   'The definitions of hazardous wastes contained in other conventions and regulations dealing in particular with the transboundary movement of hazardous wastes resemble to a large extent the definition of the Basel Convention. This applies to the Bamako and Waigani Conventions, to the Izmir Protocol to the Barcelona Convention and the Tehran Protocol to the Kuwait Convention, as well as to the OECD and EU regulations', see Jan Albers, *Responsibility and Liability in the Context of Transboundary Movements of Hazardous Wastes by Sea – Existing Rules and the 1999 Liability Protocol to the Basel Convention* (Springer 2014) 204.

contained in Annex III'.[74] In addition, the Basel Convention excludes specific waste streams from the scope of the Convention.[75] A small difference however is that the Bamako and the Waigani Conventions define hazardous waste as any substance that falls within either annex.[76] Similarly, the EU defines hazardous wastes as wastes contained in a specific list.[77]

In addition to the lists, some international instruments allow for other hazardous wastes defined as such by the domestic legislation of the party of export, import or transit to become part of the scope of the convention.[78] This approach has the advantage of creating a flexible mechanism but the disadvantage of undermining clarity in the definition of hazardous wastes under these instruments.

The definition of waste is more opaque when comparing the classification of the OECD/Basel Convention with other treaties, such as the HNS Convention[79] or the MARPOL,[80] which establish hazard definitions according to their own scope.[81] In fact, the different approaches become evident whether wastes are defined in environmental agreements, in maritime treaties, or by the EU. Discrepancies can be also found in relation to radioactive waste, which is defined as a material by the IAEA[82] and as waste by the Waigani Convention.[83] In addition, it is noticeable that international instruments divide wastes into different

---

[74]   Basel Convention (n 11) Art. 1.

[75]   Ibid., Art. 1 (3) and (4).

[76]   Waigani Convention (n 40) Art. 2, Bamako Convention (n 40) Art. 2.

[77]   EU Directive 2008/98/EC on waste and repealing certain Directives (19 November 2008) OJ L 312/3, makes reference to the EU Commission decision 2000, which provides a list of hazardous wastes.

[78]   Basel Convention (n 11) Arts 1 and 4; Bamako Convention (n 40) Art. 2; Waigani Convention (n 40) Art. 2.

[79]   HNS Convention (n 68) Art. 1 (5).

[80]   MARPOL Convention (n 38) Art. 2.

[81]   Albers (n 74) 205.

[82]   See Joint Convention on the Safety of Spent Fuel Management and on the Safety of Radioactive Waste Management (n 39) Art. 2, h:

> ... radioactive material in gaseous, liquid or solid form for which no further use is foreseen by the Contracting Party or by a natural or legal person whose decision is accepted by the Contracting party, and which is controlled as radioactive waste by a regulatory body under the legislative and regulatory framework of the Contracting Party.

[83]   Waigani Convention (n 40) Art. 1: '... wastes, which, as a result of being radioactive, are subject to other international control systems ... applying specifically to radioactive materials'.

categories such as garbage, food wastes, domestic wastes and operational wastes;[84] or wastes, hazardous wastes, waste oils and bio-waste.[85]

Other conventions, such as the Stockholm Convention on Persistent Organic Pollutants[86] and the Minamata Convention on Mercury,[87] provide definitions of wastes taking into account the specific scope of each Convention. Once these Conventions provide more clarification on the type of wastes covered by them (i.e. persistent organic pollutants as wastes, mercury wastes), this type of definition adds clarity to the scope of international conventions, such as the Basel Convention, as well as improving the national definition of waste in Parties to these Conventions.

In sum, most international instruments rely on a list to define hazardous wastes. While this approach has its advantages, it must be noted that if a specific type of waste is to be traded between two countries, it is necessary first, to identify the relevant instruments before identifying which ones are in force and applicable to the countries, including transit countries. Then, at a later stage, one must assess if the type of waste in question is defined as hazardous based on the annexes of each instrument, in addition to the national laws of each country.

It is important to observe that while international instruments provide for a rather opaque system of definition, the definition of wastes continues to evolve in order to take into account wastes intended for recycling or reclamation operations.[88] The need to continue improving the definition of waste in international law is an important challenge towards a resource-based approach to wastes.

2.4 THE ROLE OF 'SOFT LAW' IN RELATION TO WASTES

Internationally agreed obligations do not provide a sufficient and unambiguous response to address the challenges to managing waste-related issues. When it comes to the economic aspects of waste, it is important to note that a 'court or tribunal is likely to be influenced by a range of ideas and sources of information and inspiration . . . In

---

84   MARPOL Convention (n 38) Reg 1.
85   EU Directive 2008/98/EC (n 78).
86   Stockholm Convention (n 44) Art. 6.
87   Minamata Convention (n 55).
88   OECD, Decision C(2001)107/FINAL (n 47).

the context of international law they are sometimes described as "soft-law"'.[89]

In reality, information exchange between countries has proved to be an efficient way to promote the exchange of practices and international cooperation,[90] and international law has been evolving to incorporate concepts and define principles applicable to waste (3.4.1) and to facilitate the development of internationally agreed technical standards on waste issues (3.4.2).

2.4.1 The Development of Principles and Concepts Related to Waste

One of the earlier international efforts on waste management was the adoption of the 'Principles Concerning a Comprehensive Waste Management Policy' by the OECD in 1976.[91] These principles established a policy framework that would later become the general formula for some of the most used concepts present in the international standards related to waste issues. These include, for example, environment management; 'production-consumption-disposal chain'; reduction at source; re-use; recycling; and economic instruments. Furthermore, the 1976 Principles established objectives for waste management, which were divided into the establishment of inventories; the organization of waste collection; disposal centres; recycling; and public awareness. The OECD principles were later developed into an international guideline, the UNEP Guidelines and Principles for the Environmentally Sound Management of Hazardous Wastes,[92] which became one of the pillars to the Basel Convention.

In addition to materials adopted by international organizations, treaties may also put forward principles and objectives for waste management. In this context, the cornerstone principle applicable to hazardous wastes seems to be the prohibition of dumping.[93] In this

---

[89] Douglas Fisher, *Legal Reasoning in Environmental Law – A Study of Structure, Form and Language* (Edward Elgar 2013) 431.

[90] David Beede and David Bloom, 'The Economics of Municipal Solid Waste' (1995) 10 *World Bank Research Observer* 113, 140.

[91] OECD, Recommendation of the Council on a Comprehensive Waste Management Policy (28 September 1976) C(76)155/FINAL.

[92] UNEP, Guidelines and Principles for the Environmentally Sound Management of Hazardous Wastes, Decision 14/30 (17 June 1987).

[93] For example, Art. 4 of the Waigani Convention (n 40) establishes a ban on dumping of hazardous wastes and radioactive wastes at sea by reaffirming the commitments under certain international instruments; Art. 4 of the Bamako Convention establish a ban on dumping of hazardous wastes at sea, internal waters and waterways. Dumping of wastes is also prohibited under Art. 4 of the

context, the Environmentally Sound Management (ESM) of hazardous wastes, defined as 'taking all the practicable steps to ensure that
hazardous wastes or other wastes are managed in a manner which will
protect human health and the environment against the adverse effects
which may result from such wastes'[94] appears as an overarching concept
related to the promotion of international standards and referred to
directly[95] or indirectly[96] by different international instruments.[97] Other
international instruments integrate different concerns to waste management practices[98] or establish variations, such as 'radioactive waste
management'.[99]

Another concept present in international treaties is 'cleaner production',
which aims to summarize the concerns related to waste generation. This
concept is defined in the Waigani Convention as 'the conceptual and
procedural approach to production that demands that all phases of the
life-cycle of a product or process should be addressed, with the objective of
prevention or minimization of short and long-term risks to humans and to
the environment'.[100] Although this concept is not present in the text of the
Basel Convention, the idea of minimization is expressed as an obligation
of means towards the parties to the Convention.[101]

Treaties are generally evasive in relation to the economic aspect of waste-
related activities. A few exceptions are the Hong Kong Convention,[102]

Barcelona Convention; Art. 36 of the EU Directive 2008/98 (n 78); and prohibited
with exceptions under Art. 4 of the London Dumping Convention and MARPOL
Convention, Annex V.

[94]   Basel Convention (n 11) Art. 2 (8).
[95]   Stockholm Convention (n 44) Art. 6; Minamata Convention (n 55)
Art. 11 (3) (a).
[96]   ILO C170 (n 40) Art. 14.
[97]   The formula of the Basel Convention (n 11) is reproduced almost identi-
cally by Regulation (EC) 1013/2006 (n 48) Art. 2 (8).
[98]   ILO C162 (n 40) Art. 19: 'employers shall dispose of waste containing
asbestos in a manner that does not pose a health risk to the workers concerned
(. . .) or to the population in the vicinity of the enterprise'.
[99]   Joint Convention on the Safety of Spent Fuel Management and on the
Safety of Radioactive Waste Management (n 39) Art. 2(i).
[100]   Waigani Convention (n 40) Art. 1. The same concept is present in a
different manner at Art. 1(5) of the Bamako Convention (n 40).
[101]   Basel Convention (n 11) Art. 4 (2)(a).
[102]   The Hong Kong International Convention for the Safe and Environmentally
Sound Recycling of Ships (Hong Kong Convention) (adopted on 11 May 2009,
not in force) UN Doc SR/CONF/45, Art. 2 (10):

Ship Recycling means the activity of complete or partial dismantling of a ship
at a Ship Recycling Facility in order to recover components and materials for

which defines the recycling of ships, and the Basel Convention, which lists operations that may lead to resource recovery, recycling reclamation, direct re-use or alternative uses.[103] The EU identifies some economic aspects by defining recovery, re-use and recycling, and by establishing a waste hierarchy placing high importance on the economic utility of waste.[104] It has also adopted directives related to different waste streams aiming, *inter alia,* to regulate the economic use of different types of waste.[105]

It remains clear that ensuring an efficient coordinating system is of the utmost importance for the development of a coherent international waste policy that addresses environmentally sound management challenges. An improved coordination, incorporating social concerns and environmental goals, is necessary for the development of international norms.[106] In order to give further attention to the economic dimension of waste, it is also necessary to give consideration to trade and investment aspects of waste, especially when it has been noted that:

> while international trade law stipulates the importance of the unhindered transfer of wastes with the objective of promoting strong and specialized waste management industries with comparative advantages in terms of specialized know-how and technologies, as well as efficient treatment operations, international environmental law focuses particularly on the potentially polluting effects which such transboundary movements may have, and will tend to restrain cross-border waste trading.[107]

In fact, when it comes to waste trade, the WTO regime establishes a complex system based on the freedom of international trade, which may be subject to restrictions based on considerations for human health and the environment.[108] However, this system presumes that a clear distinction of wastes and non-wastes, as well as an internationally agreed definition for the different types of wastes, is in place, which presents fundamental

---

reprocessing and re-use, whilst taking care of hazardous and other materials, and includes associated operations such as storage and treatment of components and materials on site, but not their further processing or disposal in separate facilities.

[103]   Basel Convention (n 11) Annex IV.
[104]   Directive 2008/98/EC (n 78) Arts 3 and 4.
[105]   See Directive 2012/19/EU from 4 July 2012 on waste electrical and electronic equipment (WEEE) OJ L197/38; and Directive 2000/53/EC from the 21 October 2000 on end-of-life vehicles OJ L269.
[106]   OHCHR and UNEP, 'Human Rights and the Environment – Rio +20: Joint Report OHCHR and UNEP' (19 June 2012) 35.
[107]   Grosz (n 8) 509.
[108]   Ibid., 486.

difficulties in its concrete application.[109] Similarly, the absence of a clear definition may generate legal disputes.[110]

Since most countries do not possess adequate technology to manage hazardous wastes, there is a need for international facilitation of waste trade and investments. International law should assist states to 'weigh between the economic importance and potential adverse effects to health for the benefit of individuals in a manner consistent with the enjoyment of existing rights and international obligations'.[111] Thus, opaque international rules create difficulties for states in exercising their regulatory power consistently, which is often necessary for the respect of international commitments on investment protection.[112]

Although the concept and objectives of waste-related policies are becoming clearer and the economic aspect of waste is achieving its due recognition,[113] international efforts will have to provide a more comprehensive approach to waste. This approach should provide unambiguous definitions of wastes and a common understanding of the possible economic and social impact of waste throughout its management options. A common understanding will be achieved through harmonizing and improving waste management practices, which is done in great part through the adoption of international standards.

2.4.2   The Development of Technical Standards on Wastes

Technical standards are a documented synthesis of uniform technical criteria, methods, processes and practices in relation to a specific issue. While more and more international standards are provided by the private sector,[114] States have the prerogative to develop technical standards

---

[109]   Ibid., 513.

[110]   See ECJ, *Case C-9/00 Palin Granit Oy* (2002) ECR I-3533; and Joined cases *C-418/97 and C-419/97 Arco Chemie* (2000) ECR I-4475.

[111]   *Metalclad Corporation v. The United Mexican States*, ICSID Case No. ARB(AF)/97/1, Award, 40 ILM 36 (2001) para. 111.

[112]   Jorge Viñuales, *Foreign Investment and the Environment in International Law* (Cambridge University Press 2012) 248.

[113]   Cartagena Declaration (n 5) para. 6: 'We reaffirm that the safe and environmentally sound recovery of hazardous and other wastes that cannot as yet be avoided, represents an opportunity for the generation of employment, economic growth and the reduction of poverty (. . .)'.

[114]   E.g. the American Society for Testing and Materials (ASTM) developed a set of standards for treatment, recovery and reuse, sampling techniques and analysis of wastes. Similarly, the International Standard Organization (ISO) also adopted environmental standards like the ISO 14001.

through a voluntary process (i.e. recommendations) or their regulatory power (i.e. legislation). In international law, technical standards are created either through a formal process of negotiation or, more commonly, as a product of international organizations.

When it comes to technical standards on wastes, the Basel Convention Technical Guidelines on the environmentally sound management of different types of hazardous wastes are the most accomplished examples of international negotiated standards for waste management. Each Technical Guideline is negotiated and created by the Open-Ended Working Group of the Basel Convention, in conformity with a mandate established by the Conference of the Parties and relevant provisions of the Basel Convention, and subsequently adopted by the Conference of the Parties. Dozens of technical guidelines are currently available at the website of the Basel Convention, covering topics such as co-processing of hazardous wastes, the environmentally sound management of used and waste pneumatic tyres, and plastic waste.[115]

International technical standards may also become part of international law with the adoption of treaties containing specific technical guidance that are relevant to parties. One example is the Regulations for Safe and Environmentally Sound Recycling of Ships, adopted as annexed to the Hong Kong Convention.[116] The Joint Convention on the Safety of Spent Fuel Management and on the Safety of Radioactive Waste Management, which establishes very specific requirements in relation to radioactive waste,[117] is another interesting example.

As mentioned, international technical standards may also be a product of the work of international organizations. For instance, the Inter-Organization Programme for the Sound Management of Chemicals (IOMC) is currently working on the harmonization of hazard classification and labelling through the implementation of the Globally Harmonized System of Classification and Labelling of Chemicals (GHS), which will enable a standard for hazard classification and compatible labelling system for hazardous wastes. Also, the World Customs Organization works on the harmonization of codes for wastes as part of the Convention on the Harmonized Commodity

---

[115] See <www.basel.int/Implementation/TechnicalMatters/Developmentof TechnicalGuidelines/AdoptedTechnicalGuidelines/tabid/2376/Default.aspx> (last accessed 13 August 2015).

[116] Hong Kong Convention (n 103) Annex.

[117] Joint Convention on the Safety of Spent Fuel Management and on the Safety of Radioactive Waste Management (n 39).

Description and Coding System,[118] which aims to adopt a uniform interpretation on wastes subject to international trade for the use of customs
systems around the world. A more straightforward example is the work of
the International Telecommunications Union (ITU), which issues technical
recommendations related to electric and electronic wastes.[119]

One of the aims of the international efforts regarding the development
of technical standards in relation to wastes is to ensure the evolution
of ESM from a principle to an 'umbrella' technical standard on waste
management. One of the most notable attempts is being conducted by
the OECD, which developed guidance on the environmentally sound
management of wastes[120] and specific waste streams.[121] It also created six
criteria called Core Performance Elements (CPEs), which are measures
that actors involved in waste management must take to ensure that
wastes are managed in an environmentally sound manner. Another
example of the efforts to translate ESM into an international standard
is the draft framework for the environmentally sound management of
hazardous wastes and other wastes, which is currently being negotiated
under the Basel Convention. [122] This framework would serve to provide
guidance to parties on the implementation of environmentally sound
management of hazardous wastes in a systematic, consistent and comprehensive manner.

---

[118] International Convention on the Harmonized Commodity Description and
Coding System (and subsequent amendments) (adopted on 14 June 1983, entered
into force on 1 January 1988) 1035 UNTS 3.

[119] E.g. ITU Recommendation TL1100: A method to provide recycling
information of rare metals in ICT products (22 February 2012); and ITU
Recommendation TL1410: Methodology for environmental impacts of Information
and Communication Technologies goods, networks and services (8 March 2012).

[120] E.g. OECD, Guidance Manual for the Implementation of the OECD
Recommendation C(2004)100 on Environmentally Sound Management (ESM) of
Waste (2007).

[121] E.g. OECD, Technical Guidance for the Environmentally Sound
Management of Specific Waste Streams: Used and Scrap Personal Computers (18
February 2003) Doc ENV/EPOC/WGWPR(2001)3/FINAL.

[122] UNEP, Report of the Open-Ended Working Group of the Basel Convention
on the Control of Transboundary Movements of Hazardous Wastes and their
Disposal at its eighth meeting (Geneva, 25–28 September 2012) UN Doc UNEP/
CHW/OEWG.8/INF/8.

2.5 CONCLUSION

Waste is a relatively common term of broad usage that remains vague and partially undefined. As such, the legal nature of waste is a complex one and, as we have seen, can be perceived through different angles. To this inherent complexity, it is also appropriate to add a dynamic element in order to reflect the social evolution that has been reshaping the relation between human beings and waste over the necessity to rethink the use of geographic space, natural resources and hazardous substances.

International rules aimed at controlling the negative aspects of wastes, in particular the potentially hazardous effects of some types of wastes for the environment and the human health, have been put in place with relative success. Nonetheless, for a sustainable approach towards consumption and production to emerge, it is necessary that the economic aspects of wastes are not ignored. In this regard, international law needs to continue adapting in order to provide effective support for this vision to be accomplished.

3. Recycling and resource recovery under the Basel Convention: historical analysis and outlook

Pierre Portas

EXECUTIVE SUMMARY

The history of the Basel Convention still needs to be written. This chapter attempts to provide a narrative based on experience and the events of the last decades. The past 25 years have seen the rise of the Basel Convention as a key international environmental instrument which aims at reducing the export of hazardous waste and ensuring that any such waste be managed in a way to protect human health and the environment. There are two interconnected factors that explain why the Convention only partially succeeded in achieving its aims. First, trade issues came into collision with the control system of the Convention, and second, a large majority of countries parties to the Convention did not and still do not possess the capacity to manage the hazardous waste they generate in an environmentally sound way.

Throughout its history, parties made constant efforts to keep a balance between environmental protection and trade while implementing the Convention. This resulted in reducing the potential of the Convention to become a universal landmark for the environmentally sound waste management based on principles applicable to hazardous waste. As a consequence, and despite its concrete achievements, the Convention disappeared from the radar screen of politicians and became a technical instrument. The issue of recycling and recovery was never resolved in a satisfactory manner within the scope of the Convention. From an historical perspective, one could witness a loss of influence of the Basel Convention. One reason is that the parties, being preoccupied by the way the Convention would relate to trade, did not invest in exploring its potential to contribute to the emerging Green Economy movement. However, it might not be too late to face up to this new challenge.

3.1   BACKGROUND

The Basel Convention is the political response to the outrageous dumping of waste from opulent nations in poor countries, known as 'garbage imperialism'. Such dumping took on a worldwide dimension in the 1970s. Waste will follow the path of least resistance in the absence of safeguards. Governments faced with such dramatic events recognized the need for a worldwide mechanism to control the export and import especially of hazardous waste due to its potential danger to people and the environment. They were prepared to impose law on themselves to protect the victims and to restore the image of those countries where the exportation took place.

The issue of recycling and resource recovery under the Basel Convention is complex. It requires an understanding of the institutional process, the economic perspective and the rising public concern and awareness on the export and management of hazardous waste. This issue is at the heart of the tensions that prevailed while preparing and drafting the text of the Convention and during its implementation. Basically, the core point of divergence among countries concerns the question of whether an outright ban on export or a measured and balanced control of transboundary movements is the most effective way to eradicate unacceptable practices and promote those that protect both people and the environment. In the end, the economic rationale moved the discussion in its favour. The Convention's operation is focused on a global control system of transboundary movements. Export and import prohibition, and bans, are complementing this trade architecture. But it took decades to clarify the matter.

3.2   UNDERSTANDING THE ISSUE

3.2.1   Institutional Narrative

Two intergovernmental institutions played a central role in determining the philosophy and content of the Basel Convention, namely, the United Nations Environment Programme (UNEP) and the Organisation for Economic Co-operation and Development (OECD). The European Community (EC) also played an important part in the way the Convention was shaped and implemented.

3.2.1.1 The UNEP contribution

In the early 1970s, UNEP launched a programme on the development of international environmental law. The recommendations of the Ad Hoc Meeting of Senior Government Officials Expert in Environmental Law, held at Montevideo from 28 October to 6 November 1981, considered the transport, handling and disposal of toxic and hazardous waste as a priority matter and foresaw, at the global level, the preparation of guidelines, principles or conventions as appropriate.[1] The Governing Council of UNEP, based on these recommendations and the preparation of the Cairo Guidelines on the environmentally sound management of hazardous waste in 1985 (which were approved by the Council at its 14th session in 1987),[2] authorized the Executive Director to establish a legal group to work on the text of a global convention on the control of transboundary movements of hazardous waste.[3] Co-operation with OECD throughout this process was recognized as important. UNEP, in particular its Executive Director, took on the difficult and challenging task, between 1987 and 1989, to bring governments, intergovernmental bodies, industry and civil society together to prepare the text of what became the Basel Convention.[4] From the UNEP perspective, the issue of banning export from OECD countries was viewed as a necessary step to solve the problem. The importance given to the protection of developing countries from the dumping of hazardous waste was salient.

---

[1]   UNEP, Montevideo Programme for the Development and Periodic Review of Environmental Law (UNEP 1981).

[2]   UNEP, Environmental Law Guidelines and Principles No. 8: Environmentally Sound Management of Hazardous Wastes (UNEP 1987).

[3]   UNEP, Environmentally Sound Management of Hazardous Wastes, Governing Council Decision 14/30, 17 June 1987.

[4]   UNEP, Reports of the Ad Hoc Working Group of Legal and Technical Experts with a Mandate to Prepare a Global Convention on the Control of Transboundary Movements of Hazardous Wastes on its work from 1987 to 1989: Documents UNEP/WG.180/3, October 1987; UNEP/WG.182/3, February 1988; UNEP/WG.186/3, June 1988; UNEP/WG.189/3, November 1988; UNEP/WG.190/4, February 1989; UNEP/IG. 80/4, March 1989. See also UNEP, 'The Basel Convention at a Glance – Meeting Documents related to the Basel Convention' (undated) available at <www.basel.int/Portals/4/Basel%20Convention/docs/convention/bc_glance.pdf> (last accessed on 25 August 2015). For an overview of the negotiation history, see Katharina Kummer, *International Management of Hazardous Wastes* (Oxford University Press 1999) 38 ff.

3.2.1.2   The OECD contribution

Since the 1980s, both the EC and the OECD established rules to control waste movements across borders among their member countries.[5] In 1986, both the EC and the OECD amended their legal texts to apply their rules to the export of waste and hazardous waste to third countries. Through a 1985 Resolution, the OECD Council developed an international system for the effective control of transborder movements of hazardous waste, including an international agreement of a legally binding character to be concluded by the OECD member states.[6] While giving itself the legitimacy to develop such an instrument, the OECD recognized the need for a global legal framework that could be applied worldwide. The OECD subsequently co-operated with UNEP in the preparation of the global Basel Convention instead of developing its own regional legal instrument.[7] In March 1992, before the entry into force of the Basel Convention in May of the same year, the OECD Council adopted a Decision on the supervision and control of transboundary movements of waste destined for recovery operations between member countries of the OECD.[8] The emphasis of OECD was on the issue of resource recovery and efficiency.

3.2.2   The Economic Perspective

The expansion of merchandise trade worldwide, the significant decline in trade barriers and deregulation and privatization have all opened up national economies. Liberalism, free trade and financial movements are the dominant forces that shape world trends. Economic forces often dictate how waste is handled. With the globalization of the economy and finance, the issues of waste (both solutions and problems) have been dispersed all over the world through international trade. All these factors have had and still have a profound effect on waste trends; even the waste minimization approach is losing ground, eroded by the new consumerism. It is also put aside in times of financial and economic crisis.

---

[5] See OECD, Guidance Manual for the Implementation of the OECD Recommendation C(2004)100 on Environmentally Sound Management (ESM) of Waste (OECD 2007). See also Kummer, ibid., 126 ff.

[6] OECD, Council Resolution on International Co-operation concerning Transfrontier Movements of Hazardous Wastes, C(85)100 FINAL, 20 June 1985.

[7] OECD, Council Resolution on Control of Transfrontier Movements of Hazardous Wastes, C(89)1 FINAL, 30 January 1989.

[8] OECD, Council Decision on the Control of Transfrontier Movements of Wastes Destined for Recovery Operations, C(92)39/FINAL, 30 March 1992.

A globalized trade is posing challenges to government policies and market forces. The increased flow of materials across borders calls for more certainty, transparency, predictability and traceability in what moves, where it moves, how it moves and for what purpose. The growing global demand for materials has increased the trade flows and changed the international market. A large portion of resource input to industrial production is returned to the environment as waste, part of it being hazardous waste that constitutes a burden on national economies and communities. However, valuable materials contained in waste could be lost for the economy. This gave rise to concerns about the sustainability of resource use and the negative environmental impacts of production and consumption. It was therefore important and urgent to reduce these impacts through better use of resources and energy efficiency. In the 1990s, the '3Rs' (Reduce, Reuse, Recycle) concept was promoted worldwide. Subsequently other policy concepts were forged, such as the circular economy; sustainable materials management; sustainable consumption and production; and sustainable value chain. The World Trade Organization introduced discussions and negotiations on environment-related issues such as environmental goods, encompassing the issue of the trade in hazardous waste.[9]

3.2.3 Public Concern and Awareness

In 1987, hazardous waste was shipped from Italy by private business operators to Nigeria for disposal. This waste was stored in the port of Koko and began to leak, creating serious health and environmental problems. The Italian government agreed to pay the cost of returning the waste to Italy. In July 1988, two ships, the 'Karin B' and the 'Deepsea Carrier', set out to return the hazardous waste to Italy. However, because of public protest in Italy, both ships were refused entry into Italian ports. It was only in December 1988 that the hazardous waste was unloaded from the 'Karin B' and temporarily stored. The 'Deepsea Carrier' continued to be held at bay, with its crew sequestered on board, until August 1989 when the ship was finally allowed to unload in Livorno. The wanderings of the 'Karin B' and 'Deepsea Carrier' became the focus of environmental protests and the cause of growing public concern about the unscrupulous export of hazardous waste. This episode was just one among other similar cases that hit

---

[9] See also Vera Weick, 'Green Economy and sustainable development', Ch 6 of this volume. On WTO law, see Mirina Grosz, 'Transboundary movements of wastes and end-of-life goods under WTO law', Ch 5 of this volume.

many countries. Public opinion called for quick action to stop this senseless trade, stimulated by quick and dirty financial gains, and to punish those unscrupulous companies and dealers who were looking for cheap options to get rid of toxic waste. The cost of disposal in OECD countries was very high and health and environment standards were very stringent compared to those countries where the waste was imported or dumped. Trade profits were driving hazardous waste to places where: the cost of disposal was cheap; there was little or no regulation over disposal; a lack of capacity to enforce what regulation there was; low labour costs; low environmental standards; and the absence of an organized public response.[10]

Non-governmental organizations (NGOs) that were involved in the preparation of the text of the Convention were unhappy about the emphasis given to the control of transboundary movements compared with proclaiming a ban of export. Years after its entry into force, a number of such organizations continued to claim that the Basel Convention was favouring trade in hazardous waste instead of prohibiting it.[11] This perception still exists and might lead to confusion in the way the control system of the Convention operates, in particular regarding the trade of recyclables and recycled or recovered materials.

Public opinion in OECD countries put pressure on their governments to rapidly negotiate a global treaty to control international movements of hazardous waste. The media also played an active role in creating awareness about the wandering and dumping of hazardous waste.

3.3 ABOUT RECYCLING AND RECOVERY

The global demand for materials, the increasing trade flows (international and intra-regional) and the changing patterns of supply and demand have a direct impact on global and domestic resources, including waste and hazardous waste. The amount of such waste generated by economic activity is rising in line with the growing demand for raw materials, with

---

[10]   François Roelands du Vivier, *Les Vaisseaux du poison, La route des déchets toxiques* (Eds. Sang de la Terre 1988); Bill Moyers, *The Global Dumping Ground* (Lutterworth Press 1991); J. Valette and H. Spalding, *The International Trade in Wastes: A Greenpeace Inventory* (Greenpeace International 1990); Jennifer Clapp, *Toxic Exports: The Transfer of Hazardous Wastes from Rich to Poor Countries* (Cornell University Press 2001); Christoph Hilz, *The International Toxic Waste Trade* (Van Nostrand Reinhold 1992) 20–21.

[11]   See e.g. Basel Action Network, available at <www.ban.org> (last accessed on 25 August 2015).

the consequence of potentially wasting materials and energy, as well as the production of greenhouse gas. Resource recovery and efficiency can reduce the negative impacts of resource exploitation, transportation, use and disposal, and may be helpful in driving policies towards sustainable resource use and conservation. Resource recovery and efficiency policies, which promote the reuse and recycling of materials having reached the end of their useful life; policies promoting the reduction of barriers to trade; the so-called circular economy; sustainable materials management; sustainable consumption and production; the sustainable value chain; and natural resource management have direct links to the way waste and hazardous waste are generated, traded and managed. Recycling and recovery are part of a transition towards a Green Economy. It is also part of domestic policies to reduce import dependency and improve national supply conditions, including sustainable raw materials substitution.

Environmentally sound policies, as promoted by the Basel Convention, can support sustainable economic development through recycling, recovery, re-use, and other operations aiming at reducing both the use of natural resources and the quantities of hazardous waste and other waste released to the environment. Recycling or recovery is said to make a positive contribution to sustainable development in terms of reducing pressure on virgin materials, safeguarding landscapes from expanding mining activities, by reducing environmental problems and economic costs associated with the disposal of all kinds of waste or by prolonging the use of products or objects.[12] Recycling across national boundaries can bring environmental benefits where there are economies of scale so that a number of companies in different countries can share a facility, avoiding use of low-standard technologies, final disposal, or long-distance shipments. The forces that drive economic growth are the same ones that damage the environment; they also could become the same forces that help repair environmental damages. However, the economic and business environmental solutions are dependent on the fluctuations of the market or on economic downturn, and they would favour an increase in production and consumption which by themselves lead to a steady increase in the generation of hazardous waste and other waste worldwide. In some countries, export of waste for reuse, recovery or recycling represents a major foreign trade sector.[13] Such economic drivers have the consequence

---

[12]   OECD, 'Resource Productivity in the G8 and the OECD. A Report in the Framework of the Kobe 3R Action Plan' (OECD 2011) available at <www.oecd.org/env/waste/47944428.pdf> (last accessed on 25 August 2015).

[13]   Jonathan Berr, 'Recyclable materials: the U.S.'s most controversial export', *Daily Finance*, 22 April 2010.

of a more pressing demand for fair and free trade in recyclables and recycled materials. This in turn might have an impact on the macroeconomics of trade for these materials.

Efficient recycling and recovery is dependent, to a large degree, on the possibility of trading recyclables internationally, because no one country possesses the skills, capacity or infrastructure to re-use, recycle or recover the immense variety of recyclable materials. The main issue remains, nonetheless, how to effectively reverse the trend in the generation of hazardous waste and other waste. Additionally, recycling and recovery, if not properly done, like any other industrial activity that is unsound, may not only cause pollution but may also be used as a disguise for sham or illegal operations. The market is driving recyclables across borders faster than the development of policies, safeguards and legislation. In turn, such a dichotomy is the source of many of the difficulties encountered while implementing the Basel Convention. Economic actors have set the scene regarding the shape of the international trade in recyclables: policy and legislation lag behind and, internationally, a well-organized coherence of action still remains to be put in place. There is currently no level playing field at the global level. Also, tools to support environmentally sound management principles worldwide and to reduce the potential negative effects of trade in recyclables are neither sufficient nor well developed. Moving from waste to resources will not, by itself, be sufficient to avoid the undesirable effects of improper recycling or recovery, nor will it eliminate the hazardous properties of certain waste streams and the need for taking all precautionary measures.

Globally, an increasing number of governments give priority to recycling strategies. As a consequence, several of them are working towards reducing barriers to trade in recyclables or recycled materials and encouraging reuse and recycling of materials. This current trend towards establishing a loop for electronic waste and other recyclables has the consequence of increasing the international flow of used or end-of-life products, part of which is illegal or carried out on the fringe of the law. There is a significant increase in the recycling, recovery and reuse of waste worldwide with a correspondingly accelerated development of a global and intra-regional trade in recyclables and of used or end-of-life equipment. A number of governments, because of uncertainties in the characterization of electronic waste, are reluctant to impose the Basel Convention's strict control procedure on trade of hazardous electronic waste, in particular where such trade brings in revenues and generates jobs.

Transforming waste into resources can have different aspects. For instance, it may be an incentive to generate more waste; to continue business as usual; or to limit regulations that favour waste minimization.

It can also lead manufacturers to produce more goods that are recyclable, with the consequence of increasing the quantity of waste generated; increasing pollution; further depleting natural resources to produce more; increasing the energy bill; or aggravating the finances of municipalities that need to dispose of a growing quantity of waste. On the other hand, it can provide an impetus to search for means to optimize reuse, recycling or recovery of existing waste; to produce goods with less hazardous components; or to design products that are easier to recycle or to prolong the life of objects. Transforming waste into resources has three dimensions: environmental, social and economic. Each dimension contains its logic and the parameters it needs to operate that are, often, not digestible by the other two dimensions. Each dimension is looked after by different public and private entities that may not communicate with each other and conduct their own political or economic agendas. Often, if not all the time, the environmental entities, whether public or private, are marginalized compared to the power and lobby of enterprises, trade, finance and business.

3.4   TENSION IN THE BASEL CONVENTION CONTROL SYSTEM

3.4.1   How the Control System of the Basel Convention Operates

To be exempt from the Basel Convention, the exporter must prove that the waste it wants to export is not hazardous under the regime of the Convention. To do so, the exporter must first identify its waste in the list of waste contained in Annex I and further detailed in Annex VIII or IX to the Convention. Then, the exporter shall prove that its waste does not exhibit any of the hazardous characteristics described in Annex III to the Convention. Failing this and assuming the waste is identified in Annex I, the transboundary movement of the waste, characterized as hazardous, will be regulated under the Convention. This control is administratively time-consuming. The state of export, party to the Convention where the exporter is licensed, needs to approve the shipment and will seek the green light from the state of import and any state of transit. Any shipment to a non-party requires special measures and agreements. The state of export will give its authorization to the exporter once it is convinced that the hazardous waste will be managed in an environmentally sound way in the state of import. All the transactions will be done through paperwork: a notification authorizing the movement and signed by all states concerned and a movement document to ensure the traceability of the shipment until its final disposal or recycling. Control of waste contained in Annex II to the Convention applies *de facto* to two categories of waste: waste collected from households and residues arising from the incineration of household waste. Two categories of waste fall outside the scope of the Convention, namely nuclear and low-level radioactive waste (although, in certain circumstances, low-level radioactive waste arising from medical, dental or veterinary sources may fall under the scope of the Convention when mixed with household waste) as well as waste (sludge) arising from the normal operation of a ship. The control system of the Convention will apply to any waste defined as hazardous by the national or domestic regulations of a state party to the Convention. Finally, parties may not export hazardous waste to a country banning its import.[14]

3.4.2 Recycling in the Context of the Basel Convention

The discussion on transforming hazardous waste and other waste into resources has been pursued by parties in the wider context of environmentally sound management (ESM) as well as the Strategic Plan for the Basel Convention and its successor, the new Strategic Framework adopted by the Conference of the Parties in 2011.[15] Some key factors to promote ESM are to ensure the safe and sound collection, treatment, recycling or final disposal of hazardous waste, and to keep it apart from other waste streams; to avoid or prevent hazardous components in new products; to phase out outdated or prohibited chemicals; and to divert hazardous waste away from landfills. The transformation of hazardous waste into resources would then form a part of the entire lifecycle of products, including the sound and safe management of chemicals in products. When transforming hazardous waste and other waste into resources, several difficulties remain, such as waste-to-energy with issues related to emissions; recycling of hazardous waste, being an industrial operation that is likely to generate highly hazardous residues which need to be properly disposed of; and co-generation or co-processing in cement kilns, which is not environmentally neutral.

---

[14] Basel Convention, Art. 1 (Scope of the Convention), Art. 4 (General Obligations), Art. 6 (Transboundary Movement between Parties), Art. 7 (Transboundary Movement from a Party through States which are not Parties) and Art. 11 (Bilateral, Multilateral and Regional Agreements). For a detailed description of the relevant procedures, see also Juliette Voïnov Kohler, 'A paradigm shift under the Basel Convention on Hazardous Wastes', Ch 4 of this volume.

[15] Conference of the Parties to the Basel Convention, Decisions IX/3 Strategic Plan and New Strategic Framework (June 2008) and 10/2, Strategic Framework for the Implementation of the Basel Convention 2012–2021 (October 2011).

The control of transboundary movements of hazardous waste is the core of the Convention and the reason for its existence. Far from expecting that the maturity of the Convention would have been accompanied by a drastic reduction of transboundary movements, the globalization of the economy has had a significant impact in increasing such movements of waste materials considered valuable.[16] Although there is evidence that export of hazardous waste for final disposal is decreasing, the movements of used and end-of-life equipment or goods are on the rise.[17] This is putting pressure on the control system of the Basel Convention because definitions, interpretations, classification or characterization of waste, especially recyclables, are neither harmonized nor consistent among parties. As a consequence, parties are regularly uncovering what looks like a growing flow of illicit movements of all kinds of waste, particularly old electrical or electronic equipment.[18] This emerging trend will require new ways of addressing illegal traffic in addition to reinforcing the existing control mechanisms. More up-stream measures will be required to reduce the flow of illicit movements of recyclables and other waste materials transported across borders. The control of the transboundary movements of hazardous waste and other waste that potentially bears an economic value, might be captured in a tension between the necessity to protect human health and the environment from the danger or risk posed by this waste, as called for in the Convention, and the rules of international trade that would promote or facilitate its trade whether or not such movements would be required for ESM reasons.[19] Under the Convention, a party can export hazardous waste, if it does not have the capacity to deal properly with it, to another party that can ensure its ESM. Free trade in materials may not always recognize the value of the issue of the capacity

---

[16]   See for example Secretariat of the Basel Convention, 'Vital Waste Graphics 3' (2012) available at <www.envsec.org/publications/vitalwaste_br_1.pdf> (last accessed on 14 August 2015).

[17]   See for example Secretariat of the Basel Convention, 'Waste without Frontiers: Global Trends in Generation and Transboundary Movements of Hazardous Wastes and Other Wastes, Analysis of the Data from National Reporting to the Secretariat of the Basel Convention for the Years 2004 to 2006' (2010) available at <http://archive.basel.int/pub/ww-frontiers31Jan2010.pdf> (last accessed on 14 August 2015).

[18]   See for example Secretariat of the Basel Convention, 'Vital Waste Graphics 2' (2006) available at <www.grida.no/publications/vg/waste2/> (last accessed on 14 August 2015); idem, 'Illegal Traffic under the Basel Convention' (2010) available at <www.basel.int/Portals/4/Basel%20Convention/docs/pub/leaflets/leaflet-illegtraf–2010-en.pdf> (last accessed on 14 August 2015).

[19]   For a discussion of international trade rules in this context, see Grosz (n 9).

(or lack thereof) of the recipient country to manage imported hazardous waste or other waste in an environmentally sound manner. The control system of the Basel Convention could be faced with a series of practical difficulties that may hamper its effective application. It is, for instance, not uncommon to export electronic waste not as waste but as used products. Sometimes export or import of recyclables, even when exhibiting hazardous characteristics, will not be controlled under waste laws if the materials are not considered waste under the applicable legislation. The Basel Convention and regional waste management treaties as well as applicable EU and OECD legislation lay down international obligations that restrict or prohibit export or import of hazardous waste for any purpose, including for recycling or recovery. However, because there is such pressure on the industrial side for increased volumes of materials worldwide, the solutions proposed may rely on the possibility to further free the trade in recyclables and recycled materials. Such business drivers may be at odds with the situation of countries that do not possess an adequate capacity to manage their own hazardous waste in an environmentally sound manner, and therefore would not be in a position to import this type of waste for recycling or recovery without further endangering human health and the environment.

The use of environmental management systems (international standards, certification, and traceability) to complement the control system of the Convention would be useful to support on-going efforts by parties to fight and prevent environmental crimes. The dramatic consequences of illegal traffic on human health, workers' safety, security, the environment, the economy and society would require a scaling-up of awareness and action to bring this issue into the public mind and get adequate support to reverse the current trend. Business prefers flexibility while the application of the Basel Convention requires the implementation of a coherent, consistent and legally binding control mechanism. The challenge lies in the ability of parties to effectively apply the legally binding provisions of the Convention while promoting the application of the ESM principles adopted by the Conference of Parties to hazardous waste and other waste that will be transformed into resources.

3.5   HOW THE CONVENTION DEALS WITH RECYCLING AND RECOVERY

3.5.1   General Remarks

The main purpose of the Convention is to minimize both the quantity and hazard potential of waste, to reduce its transboundary movements to a minimum, and to treat and dispose of such waste as close as possible to its source of generation. The underlying principle of the Convention is the environmentally sound management of the waste. Its operational arm is a global control system of export, transit and import.

ESM contains an intrinsic ambiguity depending on the angle from which you look at it. On the one hand, it calls for restriction of trade through principles such as self-sufficiency in waste disposal, the proximity principle, and the reduction of transboundary movements.[20] On the other hand, it calls for removing barriers and other hindrances to trade, especially trade on recyclables and recycled materials, in order to save materials, resources and energy.[21] In the context of the 3Rs, as promoted by OECD,[22] the definition and classification of the materials becomes critical in terms of promoting freer movements. As a consequence, the following issues can be seen as critical:

●   waste versus non-waste issue;
●   when a waste ceases to be a waste;
●   product versus waste.

One obstacle to achieving ESM is that there is no level playing field at the global level. ESM, as a universal set of principles, needs to be applied transversally across all economic sectors, between and inside countries and between regions. Otherwise, waste will normally follow the path of least resistance and will end up where labour, social and environmental standards are lowest. One cannot rely solely on the market to protect the environment. The clarification on the waste/non-waste issue (if it can be done one day) will not, by itself, provide the required safeguard. Indeed, whether it is a product or a waste that is being dismantled, the intrinsic properties of the material will not change. If that material contains

---

[20]   See Rosemary Rayfuse, 'Principles of international environmental law applicable to waste management', Ch 1 of this volume.
[21]   See Grosz (n 9).
[22]   OECD, 'Resource Productivity' (n 12).

hazardous components and is not properly dismantled, the environmental problems will remain. This is what is happening with electronic waste today on a large scale.

The environmentally sound management principles, developed and instrumented in the context of the implementation of the Basel Convention and OECD Recommendation C(2004)100, provide the required long-term perspective and workable framework for assisting governments and industry to move towards building a recycling society respectful of people and the environment. The recycling of electronic waste, for instance, is faced with a dual constraint. On the one hand, there are international obligations that restrict or prohibit export or import of electronic hazardous waste for any purposes, including such waste destined for recycling. On the other hand, many countries do not possess an adequate capacity to recycle electronic waste. Additionally, some of the recycling plants taking imported electronic waste do not operate in a way to protect human health and the environment. A majority of countries in the world do not even possess the capacity to manage the waste they produce in an environmentally sound manner.

3.5.2 The Implementation of the Convention

The implementation of the Convention relies on the operation of two Annexes, Annex I and Annex III as referenced in Article 1 of the Convention. Annex I has been further complemented by Annexes VIII and IX. Through their experience in implementing the Convention, parties recognized that Annex I was too generic and that it was difficult to compare the list of waste with the lists produced by OECD in its regulation. Between 1999 and 2002, the Technical Working Group of the Conference of Parties worked towards harmonizing the classification system of the Convention with other systems, especially that of the OECD.[23] The way it addressed the issue was to elaborate two detailed lists of waste within the boundaries of Annex I. The Working Group quickly found that this exercise was opening up new questions, in particular the waste versus non-waste issue; it resolved the matter by creating Annex IX which contains a list of waste (List B) that will not be waste covered by Article 1 paragraph 1(a), unless it contains Annex III characteristics. Annex IX is a mirror of Annex VIII that contains a list of waste (List A) characterized as hazardous under the Convention, which is an expansion of Annex I. Both Annexes VIII and IX are used to implement

---

[23] Work of the Technical Working Group of the Basel Convention at its 15th to 20th Sessions, 1999–2002 (documents on file with the author).

the Convention; both contain waste destined for recycling and recovery. The advantage is that Annex VIII provides details of some waste to be controlled and its mirror entry in Annex IX that, normally, will not be subject to the Convention's control system.[24]

Although the addition of Annexes VIII and IX marked an operational improvement compared to the listing in Annex I, it took years for parties to come to a common agreement on how to control electronic waste, including computers, television sets or mobile phones. The issue was of definitional nature: when does a waste cease to be a waste? When does a product reach the end of its useful life? What is the meaning of reuse? These were among the key questions raised. It is through the development of partnership platforms such as the Mobile Phone Partnership Initiative (MPPI) that progress could be made in reaching consensus on definitions. The production of technical guidelines on electronic waste provided a landmark on how such materials should be dealt with within the Convention. However, there is still pressure to circumvent the Basel Convention's control system that is perceived by some economic actors as a barrier to free trade. Although, politically, parties are committed to the rules of the Convention, it is clear that the economic pressure imposes restrictions in achieving the environmentally sound management of hazardous waste worldwide.

Parties requested that the Secretariat co-operate with the World Customs Organization (WCO) in exploring the possibility of identifying separately hazardous waste in the Harmonized System (HS) of the WCO. This would have helped to improve control at borders. Once again, the economic interests represented by governments prevailed over the environmental considerations and limited the ability of the Secretariat to propose entries in the HS.[25] This was particularly true in the case of electronic waste, which was opposed by a number of governments.

---

[24]   For example, Entry A 1180 of Annex VIII reads:

Waste electrical and electronic assemblies or scrap containing components such as accumulators and other batteries included on List A, mercury-switchers, glass from cathode-ray tubes and other activated glass and PCB-capacitors, or contaminated with Annex I constituents (e.g. cadmium, mercury, lead, poly-chlorinated biphenyl) to an extend that they possess any of the characteristics contained in Annex III (note the related entry on List B B1110).

[25]   Co-operation with the WCO has been a standing agenda item of the Conference of the Parties and the Open-Ended Working Group for years. The lack of result is reflected, for example, in Decision IX/19 of the Conference of the Parties in 2008. See also <www.basel.int/Implementation/TechnicalMatters/WCOHarmonisedSystemCommittee/tabid/2390/Default.aspx> (last accessed on 14 August 2015).

The control system of the Convention might be circumvented through the conclusion of bilateral agreements (see Art. 11) between parties. Often, such agreements cover recycling or recovery issues.[26] This led to numerous discussions among parties who decided to elaborate principles on the use of such agreements.[27] According to the prevailing opinion, *in fine*, parties have a sovereign right to enter into such agreements even when there may be a question about their environmental soundness. It is up to the parties concluding the agreement to prove that such agreement is in conformity with the provisions of Article 11 of the Convention.

3.5.3   Role of the Conference of Parties

The issue of the control of recyclables was introduced by the Executive Director of UNEP who submitted a proposal to the first meeting of the Conference of Parties (COP) in Uruguay in 1992. The proposal was adopted in Decision I/16, entitled Transboundary Movements of Hazardous Wastes Destined for Recovery. This decision recognized the divergence of opinion regarding the control of hazardous waste destined for recycling and recovery and requested its Technical Working Group to make recommendations at the second meeting of the COP. In its Decision II/14, the COP, in 1994, requested the Technical Working Group to pursue its work on clarifying the issue, and to develop technical guidelines. Decision III/14 of the COP in 1995 adopted the Guidance Document on Transboundary Movements of Hazardous Waste Destined for Recovery Operations prepared by the Technical Working Group. The guidance document sets out basic principles for the environmentally sound management of waste and hazardous waste destined for recycling and recovery.

At the beginning, the control system of the Basel Convention was perceived by highly developed industrialized countries as an obstacle to free and fair trade. A number of governments claimed that the implementation of the Convention had a negative impact on the export or import of valuable materials that were needed for their economic development. For instance, trade in used lead-acid batteries is important to recover lead. Asia imported large quantities of these used goods from OECD countries. With the entry into force of the Basel Convention, this trade became part of the control system of the Convention and substantially delayed business

---

[26]   For example, the Bamako and Waigani Conventions. For other examples and a discussion of this issue, see Kummer (n 4) 87 ff.

[27]   In particular during the discussion of the Legal Working Group of the Basel Convention on the applicability of Article 11, in the first years following entry into force of the Convention.

transactions or prohibited them if it was recognized that such a trade was not environmentally sound. Under pressure to address this critical issue, the parties to the Convention initiated discussions on the issue of transforming waste into resources within the scope of the Convention. These culminated in a number of policy decisions and recommendations. The Basel Declaration on Environmentally Sound Management, adopted by Decision V/33 of the Fifth Meeting of the Conference of the Parties in December 1999, insists on the importance of prevention, minimization, recycling, recovery and disposal to achieve the environmentally sound management of hazardous waste. The Declaration was put into effect through the 2002 Strategic Plan (2002–10). The Ministerial Statement on Partnership for Meeting the Global Waste Challenge, issued at the Seventh Meeting of the Conference of Parties in October 2004, clearly stated that the challenge is to promote a fundamental shift in emphasis from remedial measures to preventive measures such as reduction at source, reuse, recycling and recovery. Another Ministerial Declaration on the Environmentally Sound Management of Electronic Waste, adopted at the occasion of the Eighth Meeting of the Conference of Parties on 1 December 2006, broke new ground. It introduced the concept of traceability, transparency and predictability in the trade of electronic waste. The MPPI represented a major step towards engaging industry in recognizing the importance of the concept of ESM as outlined by parties and of the necessity of having a strong control system for transboundary movements of hazardous waste. The Partnership for Action on Computing Equipment (PACE), adopting the same approach as the MPPI, was launched by Decision IX/9 of the Ninth Meeting of the Conference of the Parties in 2008. By Decision 10/2 in 2011, the Tenth Meeting of the Conference of the Parties adopted the Strategic Framework of the Convention (2011–21) to succeed to the Strategic Plan of Action.

An export ban amendment was first adopted as part of a decision of the Second Meeting of the Conference of the Parties in March 1994. The text of the decision provides for the prohibition by each party included in the proposed new Annex VII (parties and other states which are members of the OECD, EC, Liechtenstein) of all transboundary movements to states not included in Annex VII of hazardous wastes covered by the Convention that are intended for final disposal, and of all transboundary movements to states not included in Annex VII of hazardous wastes covered by paragraph 1 (a) of Article 1 of the Convention that are destined for reuse, recycling or recovery operations.[28] The following year, the ban

---

[28]   Decision BC-II/12, UNEP/CHW.2/30, March 1994.

was formally adopted as an amendment to the Convention.[29] Although the amendment was adopted and parties were in agreement to immediately prohibit all transboundary movements of hazardous waste that are destined for final disposal operations from OECD to non-OECD states, they could not agree when it came to prohibiting such movements destined for recycling or recovery. The 1995 Ban Amendment may not enter into force for years. Finally, at its Tenth Meeting in 2011, the Conference of Parties adopted Decision BC-10/3 to improve the effectiveness of the Basel Convention. Section A of this decision addresses the entry into force of the Ban Amendment and agreed on an interpretation of Article 17(5) of the Convention on amendments to the Convention.[30] Through this legal artefact, the parties concluded a tale that lasted for more than a decade – a point of disagreement that originated while drafting the text of the Convention; more than 20 years of tension, disagreement and acrimony that impacted negatively on the effective implementation of the Convention. It had the dramatic effect of keeping funds for implementation at a very low level not commensurate with the needs of the Convention.

3.5.4 A New Business Model: Partnerships

So-called public-private partnerships constitute a business model that has the support of industry. The Secretariat of the Basel Convention has entered into a number of partnerships to promote the ESM of waste covered by the Convention, in particular regarding used lead-acid batteries, end-of-life mobile phones and computers. These partnerships aim at improving environmental performance, monitoring and the control of export and import. Such partnerships have produced guidelines, action plans and strategies to be implemented at the domestic, regional and global level.[31] The partnerships have been useful in making industry share its expertise and in influencing the legislative process at the Convention level. It has helped make the economy

---

[29]   Art. 4A and Annex VII, adopted by Decision BC-III/1, UNEP/CHW.3/35, September 1995 (not in force).

[30]   For a full discussion see Voïnov Kohler (n 14).

[31]   Secretariat of the Basel Convention, 'Information Note on Mobile Partnership Initiatives' (undated) available at <http://archive.basel.int/pub/leaflets/leafMPPI.pdf>. Idem, 'The Partnership for Action on Computing Equipment' (undated) available at <http://archive.basel.int/industry/compartnership/>. The Guidelines are available at <www.basel.int> (all last accessed on 14 August 2015). See also Decision BC-10/19 of the Conference of the Parties, listing the relevant partnerships active in 2010.

of recycling and recovery of hazardous waste an acceptable option to achieving the aims of the Basel Convention provided it follows the ESM principles adopted by the parties. The partnerships have helped the public and private stakeholders concerned to address, within the limits of the Convention, the complex issue of the impacts of globalized trade in commodities and hazardous waste on the obligations of the Convention, and the resulting need to clarify further its scope. This has been obvious regarding the trade in old hazardous electrical and electronic equipment, for which it remains uncertain as to whether or not it may be captured within the control system of the Convention. They also looked at the complexities such trade brings into regional trade and domestic regulations.

3.6 EVOLVING TRENDS

The way the Convention can evolve is principally through the modification or addition of Annexes; and this has been the method chosen by parties in their response to emerging economic and trade issues concerning or related to recycling and recovery. Over 25 years, the parties' perception of the Convention has changed. In retrospect, one can distinguish three major steps unfolding over time. At the beginning, the urgency was to establish global control over the export and import of hazardous waste to stop irresponsible, unsound or criminal activities by putting into place the necessary legislative instrument. As a second step, parties worked together to establish international norms to ensure that the waste subject to transboundary movement under the Convention be managed in such a way as to protect human health and the environment. The third step called for developing universal norms for the environmentally sound management of waste. Logically, ESM should apply to every waste, whether hazardous or not. The Technical Working Group of the Basel Convention recognized this basic principle in its ESM guidance document approved by the Conference of Parties in 1994.[32] Since then, a number of parties have questioned this principle, fearing that it might impact on the non-hazardous waste trade. Parties have taken a number of convoluted decisions to overcome this ambiguity without resolving it. Through various decisions

---

[32] UNEP, 'Guidance Document on the Preparation of Technical Guidelines for the Environmentally Sound Management of Wastes Subject to the Basel Convention' (UNEP 1994) available at <www.basel.int/Implementation/TechnicalMatters/ DevelopmentofTechnicalGuidelines/AdoptedTechnicalGuidelines/tabid/2376/ Default.aspx> (last accessed on 13 August 2015).

and declarations, the parties called for a change of emphasis by promoting waste minimization, integrated waste management, life-cycle approach to materials or regional approach to ESM.[33] This did not really change the course of action. The enthusiasm shown in the first years of implementation of the Convention vanished rapidly, absorbed in particular by budget irritation – finance has too often prevailed over substance.

The introduction of recycling issues into the functioning of the Convention did not transform it in any depth. The control system remains the same and the ESM principles have not evolved much since 1994. The core of the Convention has remained stable. The addition or modifications of Annexes have clarified the scope of application of the Convention. There were attempts by several parties and industry, motivated by economic reasons, to initiate discussions on definitions, but these did not materialize. The Technical Working Group of the Conference of Parties responsible for developing and strengthening the technical dimension of the Convention did not engage in definitional issues. Such discussions, if successful, would have had a significant impact on the scope of the Convention, which was the intention of those who pushed for opening up this topic to exclude some recyclables from the Basel Convention control system. The very low level capacity of the majority of parties to manage hazardous waste in an environmentally sound manner acted as a conservative force to keep the Convention unchanged. Otherwise, these vulnerable countries would have lost control over the implementation of the Convention and would have been exposed to imports they could not control. One could say that the hard disc of the Convention remains, for the moment, carved in stone while its soft part (ESM) is regularly challenged by economic and trade actors.

In reality, it is the people who have changed. The drafters had the enthusiasm to build an international legal instrument to protect people and the environment from the dangers posed by hazardous waste. This did not last for long; all too quickly the difficulties inherent in implementing the treaty overcame this enthusiasm and, over the years, the Convention has become less and less attractive in the eyes of diplomats and technocrats. Additionally, in the age of the Internet, parties have not modernized the operation of the Convention, especially the way the control system functions, to adapt it to emerging economic trends or new trade patterns. The

---

[33] See e.g. the Basel Declaration on Environmentally Sound Management, adopted on the occasion of the Fifth Meeting of the Conference of the Parties to the Basel Convention and the tenth anniversary of the adoption of the Basel Convention, 6–10 December 1999.

influence of the Basel Convention has diminished and very few would now promote the Convention as a useful tool to generate green jobs. This has been a missed opportunity. The ESM of waste is part of the Green Economy both in promoting energy efficiency and in opening up new economic sectors for the recycling and recovery of hazardous waste generated in countries where such activities are at a low level. Parties did not explore ways and means to make the Convention an asset for building a Green Economy. Is it too late? To move in that direction, it would be important to increase the political visibility of the Convention and articulate its potential to contribute to the greening of trade and the economy.

3.7   A LIMPING PROGRESS

3.7.1   A Gap Between Vision and Implementation

The first decade of the implementation of the Basel Convention was devoted to clarifying and improving its operation. The Secretariat played an active and crucial role in helping and guiding parties in this process. The second decade, as agreed at the Fifth Meeting of the Conference of the Parties in 1999, was to make ESM accessible to all parties, with an emphasis on the minimization of waste covered by the Convention and the development of capacity building. However, the parties never provided the Convention with the means to achieve such ambitious objectives. One reason was that the parties were of the opinion that the Secretariat should not be involved in implementation, although there was no other body capable of replacing the Secretariat in this task. This had an impact on the ability of the Secretariat to carry out concrete action on the ground. The establishment of the regional and co-ordinating centres within the Convention did not eliminate this intrinsic weakness of the Convention, which is neither supported by a global programme on the ESM of waste and hazardous waste nor by a financial mechanism. Each conference of the parties adopted new decisions imposing further work on the Secretariat while keeping the level of funding to the minimum possible. As a result, the Secretariat could only launch a limited number of concrete actions regarding recycling and recovery. The issue of Annex VII (Ban Amendment) further complicated the matter as it built a psychological and political barrier which prevented the Secretariat from testing case studies on recycling and recovery. At the end of the day, one could sum up the situation by saying that the vision of the parties was supported by a limping Secretariat and impoverished by a profound divergence among parties: those who promoted trade in recyclables and those who resisted such trade. Parties never gave a clear and strong direction, due to these circumstances.

3.7.2 A Paradox and a Collision

Securing prosperity, economic development and the sustainable use of natural resources, which depend on a reliable supply of both primary and secondary raw materials, impacted on the implementation of the Convention, especially with regard to achieving self-sufficiency in waste management and waste minimization: as trade takes place across borders, transboundary movements are often necessary for ensuring reliable material supply. Furthermore, the boost in emerging technologies and the transformation of energy systems (waste-to-energy) defeated the purpose of reducing movements of hazardous waste across frontiers, as many such operations by necessity involve transboundary movement.[34] The underlying conflict between market forces and the regulatory framework of the Convention have collided throughout the history of the treaty. The philosophy of a social order based on freedom given by a competitive open market where self-interest drives the economy influenced parties to the Convention to bend towards a preference for a treaty that focuses on the control of transboundary movements of hazardous waste, rather than the foundation for a global programme on the ESM of waste.

3.7.3 Limping Pillars

There are two pillars that constitute the Basel Convention, namely its control system and the ESM principles underlying its provisions. Four main policy directions were agreed by parties to move into the second decade of the implementation of the Convention: waste minimization, integrated waste management, life-cycle approach to materials and a regional approach.[35] This led to new considerations by parties on the relationship between the implementation of the Convention and trade. Issues debated in recent times have included ship dismantling or scrapping; classification of, and control systems for, used and scrap electronics; and materials for repair or refurbishment or remanufacturing. Some governments, parties and signatories, considered that the current Basel

---

[34] For example, importation of used tyres into a country for the purpose of producing energy, or the establishment of cement kilns in developing countries that will require waste imported from within the region in order to be productive.
[35] Basel Declaration (n 33).

Convention system for controlling international shipments of hazardous waste makes trade in many of these materials difficult and in some cases impossible, and supported consideration of alternative systems of control under the Convention. Other parties argued that the Convention should apply, in its current form, to the international movement of used products for repair, refurbishment, or remanufacture. In response, some parties' position has been that international movement of equipment for repair, refurbishment, or remanufacturing does not constitute movement of waste, and thus is not impacted by the Convention or its procedures. This divergence on the scope of the Convention might have an effect on the fulfilment of ESM for hazardous recyclables, in particular within countries that do not possess and are unlikely to possess in the near future the capacity for ESM.

3.8   CONCLUDING REMARKS

This short voyage through the Basel Convention's past illustrates the dichotomy that prevailed among those who built, developed and implemented the treaty. This dichotomy is the mirror of the tensions that move inside society. The economy and the environment are at odds. The Basel Convention today should be part of the push for a Green Economy through the promotion of energy efficiency, sound recycling and the sustainable use of materials. But, in reality, it is still perceived like sand in your shoes. With time, its authority might be eroded in a world where liberalism remains the dominant attractive way of trading goods. Parties to the Convention, *de facto*, became schizophrenic. They were responding to their citizens' claim for a robust control of the transboundary movements of hazardous waste while, at the same time, promising the same people more jobs and economic growth based on trade, including trade in hazardous waste. The Convention never could fulfil its potential. There was always a reason to water down its impact by opening new fronts like definitions or classification. The proposed Ban Amendment offered an objective reason to many parties to limit funding to the Convention, arguing that a ban would not be in favour of ESM.

Those who drafted the text of the Convention were inspired by the necessity to stop the infamous trade of hazardous wastes that were dumped in vulnerable countries. The Basel Convention has partially achieved this ambitious goal. The Convention was not designed in such way as to build an intrinsic flexibility for accommodating trade issues within its control system. The only way it could absorb such a disruptive force was through the amendments of its Annexes or the addition of new

Annexes. The challenge was to keep a coherence between its overall goals and its implementation. The story is not finished yet. Pressure on the control system might continue and the ESM principles might be watered down. The issue is not of a technical but of a political nature.

Can we argue that the Basel Convention control system has finally absorbed the complex issue of recycling and recovery? Tomorrow, some parties might lay down new arguments to exclude some hazardous recyclables from the scope of the Convention. Clearly, there should be a point where the core of the Convention should be preserved over time; otherwise, the Convention might look like a tennis racket where there are more holes than matter. The effective and efficient implementation of the Basel Convention relies on trust among the public and private stakeholders; and honesty and transparency in what moves across borders, where and how it goes and how it is transported and disposed of. At the end of the day, the Convention is what parties make of it.

4. A paradigm shift under the Basel Convention on Hazardous Wastes

Juliette Voïnov Kohler[1]

EXECUTIVE SUMMARY

The 1989 Basel Convention aims to protect human health and the environment against the negative impacts of hazardous and other wastes. Although a pre-Rio treaty, the Convention is not oblivious to social and economic concerns and contains the necessary provisions to ensure that such considerations are taken into account when achieving its environmental objective. The Basel Convention is based on a life-cycle approach: it sets out obligations pertaining to the generation of wastes and to the management of wastes, including their transboundary movements. Over the years, the parties to the Convention have given concrete meaning to the obligation to ensure the environmentally sound management of wastes. They have also striven to strengthen the treaty's trade control regime through the adoption, in 1995, of a ban on the export of wastes from developed to developing countries. Less emphasis however was directed to the reduction of waste generation. During the Ninth Meeting of the Conference of the Parties in 2009, a decisive political push by the Indonesian President of the Conference of the Parties, relayed by Switzerland through the Country-Led Initiative, opened the door to overcoming the long-standing political deadlock over the ban. Colombia, in its capacity as host of the Tenth Meeting of the Conference of the Parties held in 2011, complemented the initiative by proposing the adoption of a Declaration on the Prevention, Minimization and Recovery of Hazardous Wastes and Other Wastes. This combination of efforts led to the historical outcomes of the Tenth Meeting of the Conference of the Parties. The meeting witnessed a paradigm shift in the Basel Convention, including the recognition of the economic potential of the environmentally sound

---

[1] The views expressed herein are those of the author and do not necessarily reflect the views of the United Nations.

recovery of wastes. In doing so, the parties to the Basel Convention gave concrete meaning to the Green Economy, a new strategic direction subsequently embraced at the Rio+20 Summit.

4.1 INTRODUCTION

The June 2012 Rio+20 Summit consecrated the Green Economy at the broadest and highest political level. The endorsement of this new strategic direction, which is to support the objectives of sustainable development and poverty eradication, was the culmination of a variety of building blocks over the years, one of which was the Tenth Meeting of the Conference of the Parties to the Basel Convention on the Control of Transboundary Movements of Hazardous Wastes and their Disposal (hereinafter, the Basel Convention).[2] Held in Cartagena, Colombia, in October 2011, this meeting can also be seen as a landmark event in the life of the Convention itself, which continues to frame its future to this day.

This chapter focuses on the normative aspects of the Basel Convention, the only global multilateral environmental agreement dealing with wastes that have the potential to harm human health and the environment. The first part of the chapter looks at the broader context, in particular how other multilateral environmental agreements contemporary to the Basel Convention have sown the seeds of sustainable development and, ultimately, Green Economy policies. The second part of the chapter introduces the Basel Convention and its key provisions, followed by a third part presenting an overview of the evolution of this treaty over the last 20 years. The fourth part of the chapter focuses on developments since the Ninth Meeting of the Conference of the Parties, which led to the historical outcomes of the Tenth Meeting of the Conference of the Parties. These outcomes and the newly found balance between the rich social, environmental and economic benefits of the Convention are set out in the fifth and final part of the chapter.

---

[2] Basel Convention on the Control of Transboundary Movements of Hazardous Wastes and their Disposal (Basel Convention) (adopted on 22 March 1989, entered into force on 5 May 1992) 1673 UNTS 57.

4.2 THE SEEDS OF A GREEN ECONOMY IN PRE-RIO MULTILATERAL ENVIRONMENTAL AGREEMENTS

The Basel Convention is, alongside a handful of global multilateral environmental agreements such as the Montreal Protocol on Substances that Deplete the Ozone Layer (Montreal Protocol),[3] the Convention on the International Trade of Endangered Species of Wild Fauna and Flora (CITES),[4] the Convention on the Conservation of Migratory Species of Wild Animals (CMS)[5] and the Convention on Wetlands of International Importance (Ramsar),[6] a pre-Rio treaty, in the sense that it was adopted prior to the 1992 United Nations Conference on Environment and Development (UNCED) that culminated in the advent of sustainable development.[7] Premised on the importance of ensuring inter and intra-generational equity, sustainable development is based on the conviction that economic development, social development, and environmental protection are interdependent and mutually reinforcing. This does not, however, mean that the Basel Convention and other pre-Rio multilateral environmental agreements are oblivious to social or economic considerations, quite the contrary.

Adopted in 1987, the Montreal Protocol already made express reference to the need to protect human health alongside the objective of protecting the environment.[8] The Protocol also acknowledges, although in a limited way, considerations of intra-generational equity in its Article 9 pertaining to research and development, by requiring that all parties cooperate towards the development of the best technologies for improving the containment, recovery, recycling, or destruction of controlled substances, or otherwise reducing their emissions. By also requiring that all parties

---

[3]   Montreal Protocol on Substances that Deplete the Ozone Layer (adopted on 16 September 1987, entered into force on 1 January 1989) 1522 UNTS 3.

[4]   Convention on the International Trade of Endangered Species of Wild Fauna and Flora (adopted on 3 March 1973, entered into force on 1 July 1975) 993 UNTS 243.

[5]   Convention on the Conservation of Migratory Species of Wild Animals (adopted on 23 June 1979, entered into force on 1 November 1983) 1651 UNTS 133.

[6]   Convention on Wetlands of International Importance (adopted on 2 February 1971, entered into force on 21 December 1975) 996 UNTS 246.

[7]   Rio Declaration on Environment and Development, Report of the United Nations Conference on Environment and Development, I (1992) UN Doc A/ CONF.151/26; (1992) 31 ILM 874.

[8]   Montreal Protocol (n 3) Preamble, para. 2.

cooperate towards the identification of possible alternatives to controlled substances, to products containing such substances, and to products manufactured with them, the Protocol also highlights the potential for business and industry to contribute to solving the problem. Taken together, these provisions of the original text of the Montreal Protocol illustrate an early endorsement of sustainable development and the precepts of a Green Economy. Social and economic considerations are also an integral part of the pre-Rio multilateral environmental agreements dealing with fauna, flora or ecosystems, in line with Principle 4 of the Stockholm Declaration on the Human Environment that provides that '(n)ature conservation, including wildlife, must therefore receive importance in planning for economic development'.[9] In the preamble of CITES, for instance, the contracting States state that they are '(c)onscious of the ever-growing value of wild fauna and flora from aesthetic, scientific, cultural, recreational and economic points of view'. With respect to wetlands, the contracting parties of the Ramsar Convention stipulate, in the preamble, that they are 'convinced that wetlands constitute a resource of great economic, cultural, scientific and recreational value, the loss of which would be irreparable'. The concept of wise use, fundamental to this treaty, was subsequently recognized as a contribution towards achieving sustainable development throughout the world.[10]

The Basel Convention, adopted in 1989, also integrates provisions that pave the way to the achievement of sustainable development and a Green Economy. For instance, its preamble places on equal footing concerns about protecting human health and about protecting the environment.[11] The preamble also acknowledges the importance of intra-generational equity by recognizing the special needs of developing countries, namely their limited capabilities to manage hazardous wastes and other wastes, the associated increasing desire for the prohibition of transboundary movements of hazardous wastes and their disposal in developing countries,[12] and the need to promote the transfer of technology for the sound management of hazardous wastes and other wastes produced locally, particularly to developing countries.[13] In addition to the integration of social concerns

---

[9]   The Declaration was subsequently adopted by the General Assembly in its Resolution 2994 (XXVII).

[10] See the Ramsar Handbook 1, 'Wise use of wetlands', available at <www.ramsar.org/sites/default/files/documents/library/hbk4–01.pdf> (last accessed on 11 August 2015).

[11] Basel Convention (n 2) Preamble, paras 1–4, 9, 14, 15 and 24.

[12] Ibid., para. 7.

[13] Ibid., para. 18.

and of considerations of intra-generational equity, the preamble hints at the potential role of business and industry in developing environmentally sound low-waste technologies, recycling options, good house-keeping and management systems with a view to reducing to a minimum the generation of hazardous wastes and other wastes.[14] Preambular text, however, has a limited legal impact unless its content is reflected in the operational part of the treaty. In this regard, one provision, Article 10 on 'international cooperation', merits special attention since it requires all parties to cooperate in the development and implementation of technologies with the potential of minimizing the generation of hazardous wastes and other wastes and ensuring their management in an environmentally sound manner. All parties must also cooperate in the transfer of technology and management systems related to the environmentally sound management of hazardous wastes and other wastes. These obligations are reminiscent of those embedded in Article 9 of the Montreal Protocol.

Despite the integration of economic and social concerns, pre-Rio treaties, in comparison to post-Rio treaties such as the United Nations Framework Convention on Climate Change (UNFCCC),[15] the Convention on Biological Diversity,[16] the United Nations Convention to Combat Desertification[17] or the Stockholm Convention on Persistent Organic Pollutants,[18] do not give equal prominence to economic, social and environmental objectives in their operational part. One notable exception is the Montreal Protocol that actually embraced the pillars of sustainable development and considerations of inter and intra-generational equity prior to UNCED through the adoption of the 1990 London amendment.[19] This amendment to the preamble and operational part of the treaty can be seen as having upgraded the protocol to a post-UNCED type of treaty, one that not only recognizes the special needs of developing countries by setting out differentiated targets and timetables for control

---

[14] Ibid., para. 21.
[15] United Nations Framework Convention on Climate Change (adopted on 9 May 1992, entered into force on 21 March 1994) 1771 UNTS 107.
[16] Convention on Biological Diversity (adopted on 5 June 1992, entered into force on 29 December 1993) 1760 UNTS 79.
[17] United Nations Convention to Combat Desertification (adopted on 14 October 1994, entered into force on 26 December 1996) 1954 UNTS 3.
[18] Stockholm Convention on Persistent Organic Pollutants (adopted on 22 May 2001, entered into force on 17 May 2004) 2256 UNTS 119.
[19] The amendment to the Montreal Protocol was agreed by the second meeting of the parties (London, 27–29 June 1990). It is available at <http://ozone.unep.org/new_site/en/Treaties/treaties_decisions-hb.php?dec_id_anx_auto=780> (last accessed on 11 August 2015).

measures and establishing a financial mechanism for the benefit of developing countries, but also paves the way for a Green Economy through the adoption of its new Article 10A on transfer of technology whereby each party is to ensure that best available, environmentally safe substitutes and related technologies are expeditiously transferred to developing-country parties.

The London amendment did set a powerful precedent for UNCED, but its impact on other pre-Rio treaties appears to have been limited. Within the Basel Convention in particular, the ultimate push for a new balance between the three pillars of sustainable development and a clearer shift towards the Green Economy was to take an additional two decades.

4.3   THE BASEL CONVENTION: AN OVERVIEW

The Basel Convention was negotiated in the wake of the ever-increasing generation of hazardous wastes and the growing enactment of more stringent regulatory frameworks in developed countries pertaining to their disposal. In search of cheaper alternatives, hazardous wastes, then essentially seen as unwanted by-products of certain industrial activities and consumerism, were shipped from developed economies to developing countries entirely lacking adequate disposal facilities to manage the wastes in an environmentally sound manner. Several incidents involving the dumping of these wastes in developing countries were brought to light,[20] leading first to the development of the 1987 Cairo Guidelines and Principles for the Environmentally Sound Management of Hazardous Wastes,[21] and subsequently to the negotiation of what would become the only global environmental agreement to date focusing on wastes posing a threat to human health and the environment. Adopted in 1989, the Basel Convention entered into force in May 1992. As of 1 March 2016, it had 183 parties, making it a nearly universal treaty.

---

[20]   One example of such an incident is the 1988 disaster in Koko, Nigeria. For more information, see Francis Adeola, 'Environmental Injustice and Human Rights Abuse: The States, MNCs, and Repression of Minority Groups in the World System' (2001) 8(1) *Human Ecology Review* 39, 50. Later incidents that have had a decisive impact on the Convention include the Probo Koala incident in 2006. For more information, see Olanrewaju Fagbohun, 'The Regulation of Transboundary Shipments of Hazardous Waste: A Case Study of the Dumping of Toxic Waste in Abidjan, Cote d'Ivoire' (2007) 37(3) *Hong Kong Law Journal* 831, 841.

[21]   The guidelines were adopted during the 14th session of the UNEP Governing Council by its Decision 14/30.

The Basel Convention aims at protecting human health and the environment against the negative impacts of hazardous and so-called 'other' wastes. Wastes are defined as 'substances or objects which are disposed of or are intended to be disposed of or are required to be disposed of by the provisions of national law' (Art. 2, para. 1). It is therefore what happens, is to happen or must happen to the substance or object that is decisive in determining its nature as 'waste'. 'Disposal' means any operation specified in Annex IV to the Convention and includes both operations that are final and operations that may lead to resource recovery, recycling, reclamation, direct reuse or alternative uses. Hazardous wastes are listed in Annexes I and VIII of the Convention and are defined as waste streams or as wastes having specific constituents. To be considered 'hazardous wastes', the wastes must meet the 'hazardous' characteristics specified in Annex III to the Convention, for instance be explosive, poisonous, infectious, toxic, flammable or corrosive. A party to the Convention has the prerogative to extend the scope of the hazardous wastes covered by the Convention by defining such wastes nationally and notifying all parties of these definitions through the Secretariat of the Convention. The second category of wastes covered by the Convention, namely 'other' wastes, is defined in Annex II: it includes wastes collected from households as well as residues arising from the incineration of household wastes.

The Basel Convention is based on a life-cycle approach: it sets out obligations pertaining to the generation of wastes and to the management of wastes, including their transboundary movements. However, the extent of the obligations undertaken by parties differs widely within this cycle. With respect to the generation of wastes, the minimization of which constitutes the first pillar of the Convention, the preamble does note that the most effective way of protecting human health and the environment from the dangers posed by such wastes is the reduction of their generation to a minimum in terms of quantity and/or hazard potential. However the main provision of the Convention pertaining to the generation of wastes, set out in its Article 4, paragraph 2, only provides that: 'Each Party shall take the appropriate measures to: (. . .) ensure that the generation of hazardous wastes and other wastes within it is reduced to a minimum, taking into account social, technological and economic aspects.'

Other provisions of the Convention set out ancillary obligations to support the 'soft law' obligation pertaining to the reduction of the generation of hazardous and other wastes, namely the obligation to cooperate in the development and implementation of new environmentally sound low-waste technologies and the improvement of existing technologies (Art. 10), and the obligation to exchange information on the effects on human health and the environment of the generation of hazardous or other wastes, and on measures undertaken for the development of technologies for the reduction and/or elimination of production of hazardous and other wastes (Art. 13, para. 3). However, in comparison to other multilateral environmental agreements setting out obligations aimed at eliminating a specific hazard, such as the Stockholm Convention, the Montreal Protocol or the Kyoto Protocol to the UNFCCC,[22] the obligations of the Basel Convention pertaining to the generation of hazardous and other wastes can be seen as relatively modest.

The environmentally sound management (ESM) of wastes is the second pillar of the Convention. ESM is defined as taking all practicable steps to ensure that hazardous wastes or other wastes are collected, transported, and disposed of in a manner that will protect human health and the environment against the adverse effects which may result from such wastes (Art. 2). The ESM requirement is further elaborated through various obligations, for instance the obligation of each party to ensure the availability of adequate disposal facilities for the ESM of hazardous and other wastes that shall be located, to the extent possible, within it; the obligation to ensure that persons involved in the management of hazardous or other wastes within it take such steps as are necessary to prevent pollution due to hazardous and other wastes arising from such management and, if such pollution occurs, to minimize the consequences thereof for human health and the environment; and the obligation to prohibit all persons under its national jurisdiction from transporting or disposing of hazardous or other wastes unless such persons are authorized or allowed to perform such functions. In order to clarify the content of the ESM requirement with respect to specific waste streams, waste constituents, hazardous characteristics or disposal operations, the Convention provides for the subsequent development of 'technical guidelines' (Art. 4, para. 8).[23]

In terms of the extent of the obligations undertaken by parties, it is the regime established to control transboundary movements of hazardous and other wastes that forms the backbone of the Convention. The Convention sets out both specific conditions for such transboundary movements to be allowed to take place and a detailed procedure that needs to be followed for each proposed movement. With respect to the conditions, one can

---

[22]   Kyoto Protocol to the United Nations Framework Convention on Climate Change (adopted on 11 December 1997, entered into force on 16 February 2005) 2303 UNTS 148.

[23]   The technical guidelines are available on the website of the Convention at <www.basel.int/Implementation/TechnicalMatters/DevelopmentofTechnical Guidelines/AdoptedTechnicalGuidelines/tabid/2376/Default.aspx> (last accessed on 11 August 2015).

mention for instance the fact that the export of hazardous or other wastes to a state which has prohibited by its legislation all imports, or to a state in which there is reason to believe that the wastes in question will not be managed in an environmentally sound manner, is prohibited (Art. 4, para. 2e). Another condition set by the Convention is the general prohibition of allowing wastes within its scope to be exported to a non-party or to be imported from a non-party, unless a specific agreement is in place that does not derogate from the ESM of hazardous and other wastes as required by the Convention (Art. 4, para. 5 and Art. 11). A third example is the prohibition on exporting hazardous or other wastes for disposal within the area south of 60° South latitude (Art. 4, para. 6). In addition to such conditions, the Convention sets out in its Article 6 a four-step procedure that needs to be followed for each proposed transboundary movement of hazardous or other wastes.

Taken together, these four steps are usually referred to as the 'prior informed consent' (PIC) procedure:

- Step 1: a transboundary movement of wastes is proposed. This proposal must be preceded by the conclusion of a contract between the exporter or generator and the importer or disposer specifying the ESM of the wastes to be moved. If all the conditions for a proposed movement are met, a notification of the proposed movement is sent by the state of export, or by the generator or exporter through the state of export, to the state of import and any state of transit.
- Step 2: consent to the proposed movement. Upon reception of the notification, the state of import and any state of transit have the possibility to consent to, with or without conditions, or to refuse the proposed movement. Their decision is to be notified in writing to the state of export, or to the generator or exporter through the state of export. A proposed movement cannot be initiated until the required consents have been received in writing.
- Step 3: the movement takes place. The state of export or the exporter issues a movement document that will accompany the shipment until the wastes are disposed of. The movement document must be signed by any person that takes charge of the shipment.
- Step 4: ESM of the wastes. The importer or disposer must confirm reception of the wastes and of their subsequent disposal in an environmentally sound manner.

The Basel Convention therefore does not ban but strictly controls, the export, transit and import of hazardous and other wastes; any state of import or transit may refuse to consent to a proposed movement of wastes. In addition, any party also has the possibility, within its national legal framework or through a regional agreement, to prohibit or restrict the export and or import of hazardous and other wastes, a possibility of which several parties have made use.[24] Reflecting the importance of complying with the control measures for transboundary movements, the Basel Convention requires all parties to consider as criminal the illegal traffic of hazardous wastes and other wastes (Art. 4, para. 3).[25]

An overview of the transboundary movements presumably taking place in accordance with the Basel Convention, as reported by parties over the period 2004–06, shows that transboundary movements over that period took place between 128 countries and involved more than 10 million tonnes of hazardous and other wastes.[26] Transboundary movements of wastes are thus a truly global activity that takes place both among developed and developing countries and between countries from either group.

4.4 THE EVOLUTION OF THE BASEL CONVENTION FROM 1992 TO 2009

Following the adoption of the Basel Convention, two regional agreements focusing on similar issues, yet with a scope extended to nuclear wastes and integrating an import ban, were adopted: the 1991 Bamako Convention on the Ban of the Import into Africa and the Control of Transboundary Movement and Management of Hazardous Wastes within Africa,[27] and the 1995 Convention to Ban the Importation into

---

[24] See the information available on the website of the Convention at <www. basel.int/Countries/ImportExportRestrictions/tabid/1481/Default.aspx> (last accessed on 11 August 2015).

[25] 'Illegal traffic' is defined as any transboundary movement undertaken without notification pursuant to the provisions of this Convention to all states concerned; or without the consent pursuant to the provisions of this Convention of a state concerned; or with consent obtained from states concerned through falsification, misrepresentation or fraud; or that does not conform in a material way with the documents; or that results in deliberate disposal (e.g. dumping) of hazardous wastes or other wastes in contravention of this Convention and of general principles of international law (Art. 9).

[26] Secretariat of the Basel Convention, 'Waste Without Frontiers' (2010) available at <www.basel.int/Portals/4/Basel%20Convention/docs/pub/ww-frontiers26 Jan2010.pdf> (last accessed on 11 August 2015).

[27] Bamako Convention on the Ban of the Import into Africa and the Control of Transboundary Movement and Management of Hazardous Wastes within Africa (adopted on 30 January 1991, entered into force on 22 April 1998) 2101 UNTS 242.

Forum Island Countries of Hazardous and Radioactive Wastes and to Control the Transboundary Movement and Management of Hazardous Wastes within the South Pacific Region (also referred to as the Waigani Convention).[28] At global level, renewed efforts to strengthen the provisions of the Basel Convention pertaining to the transboundary movements of hazardous wastes took place as early as the First Meeting of the Conference of the Parties in 1992 with a call by developing countries for a ban on all exports of hazardous wastes from country members of the Organisation for Economic Co-operation and Development (OECD) to non-OECD countries. This call was intended to address challenges faced by the latter in controlling imports of hazardous and other wastes they were unable to manage in an environmentally sound manner. The proposal was concretized by the adoption of Decision II/12 during the Second Meeting of the Conference of the Parties in 1994, subsequently adopted as an amendment to the Convention at the Third Meeting of the Conference of the Parties in 1995 as a new Article 4A (Decision III/1). Although it was adopted by consensus, several delegations expressed their reservations with respect to the amendment, in particular Australia, Canada, New Zealand and the Russian Federation,[29] thereby signalling some level of discomfort with it.

In accordance with this so-called Ban Amendment, each party listed in a new Annex VII (comprising the members of the OECD, of the European Union, and Liechtenstein) would be required to prohibit immediately all transboundary movements of hazardous wastes destined for final disposal operations to states not listed in Annex VII, and to prohibit as of 31 December 1997 all transboundary movements of hazardous wastes destined for recovery or recycling operations to such states.[30] By the time of the Ninth Meeting of the Conference of the Parties (which took place in Indonesia in 2008), parties were still in disagreement over the interpretation of Article 17 paragraph 5 of the Convention pertaining to the required threshold for the entry into force of amendments to the Convention,

---

[28]   Convention to Ban the Importation into Forum Island Countries of Hazardous and Radioactive Wastes and to Control the Transboundary Movement and Management of Hazardous Wastes within the South Pacific Region (adopted on 16 September 1995, entered into force on 21 October 2001) 2161 UNTS 93.

[29]   See the Report of the Third Meeting of the Conference of the Parties, paragraph 51 and Annexes I, II and III, available at <www.basel.int/TheConvention/ ConferenceoftheParties/ReportsandDecisions/tabid/3303/Default.aspx>(lastaccessed on 11 August 2015).

[30]   See   <www.basel.int/Implementation/LegalMatters/BanAmendment/ tabid/1484/Default.aspx> (last accessed on 11 August 2015).

including the Ban Amendment.[31] Although framed as a legal issue, the various views put forward on the matter reflected essentially a political disagreement over the Ban Amendment itself.

In parallel to these efforts to strengthen the control regime pertaining to the transboundary movements of hazardous wastes, the parties to the Convention launched a series of negotiations of technical guidelines aimed at clarifying the obligations of parties with respect to ensuring the environmentally sound management of wastes, as noted above. Adopted by the Conference of the Parties, these guidelines are not, per se, legally binding. However, as they reflect the global technical and political consensus on the meaning of ESM, technical guidelines have considerable weight. It is worth noting that these guidelines are developed with the understanding that waste management is designed to identify and manage wastes throughout their entire life cycle and that waste management should rely on the following waste management hierarchy: waste avoidance/minimization; recovery; final disposal. In other terms, where waste avoidance is not possible, then reuse, recycling and recovery, where possible, become the preferable alternative to final disposal. An impressive number of guidelines have been adopted over the years, and work is ongoing.[32]

With respect to the generation of wastes, one may observe from decisions adopted over the years by the Conference of the Parties that less emphasis was directly placed on strengthening Article 4 paragraph 2 of the Convention, which mandates reduction of waste generation. During its Fifth Meeting in 1999, on the occasion of the tenth anniversary of the adoption of the Basel Convention, the Conference of the Parties adopted a ministerial declaration on ESM which, among other things, recognizes that, notwithstanding the concerted efforts made during the first decade of the Basel Convention, hazardous waste generation had continued to grow at global level. In the declaration, the parties reaffirm that the prevention and minimization of the generation of hazardous and other wastes are fundamental aims of the Convention. This message was taken up again at the Seventh Meeting of the Conference of the Parties in 2004 through the adoption of Decision VII/2 entitled 'Hazardous Waste Minimization' which called upon all parties and other states to increase their efforts to

---

[31] For the various interpretations put forward, see Decision IX/25 available in the Report of the Ninth Meeting of the Conference of the Parties, available at <www.basel.int/TheConvention/ConferenceoftheParties/ReportsandDecisions/tabid/3303/Default.aspx> (last accessed on 11 August 2015).

[32] See the website of the Basel Convention, <www.basel.int/Implementation/TechnicalMatters/DevelopmentofTechnicalGuidelines/tabid/2374/Default.aspx> (last accessed on 11 August 2015).

take steps to reduce the generation of hazardous wastes and other wastes subject to the Basel Convention and to share their experiences in this respect. Only one party, namely Norway, had submitted information by the time of the Eighth Meeting of the Conference of the Parties,[33] at which time, by its Decision VIII/23, the Conference of the Parties recalled its Decision VII/2 and once again invited parties and others to provide comments to the Secretariat by 30 June 2008 on their experiences with hazardous waste minimization. No such submissions were received by 30 June 2008, and no further decision was adopted specifically on this matter.[34]

4.5   THE NINTH MEETING OF THE CONFERENCE OF THE PARTIES: PAVING THE WAY TOWARDS THE 2011 HISTORICAL OUTCOMES

The Ninth Meeting of the Conference of the Parties which took place in 2008 in Indonesia, and in particular the President's statement on the possible way forward on the Ban Amendment,[35] was decisive in shaping the historical agreements later reached during the Tenth Meeting of the Conference of the Parties. A key element of this statement is the affirmation of the objective of the Ban Amendment, seen as a mechanism to protect vulnerable countries without adequate capacity to manage hazardous wastes in an environmentally sound manner, and to ensure the ESM of hazardous wastes. Placing ESM, which can be seen as the ultimate objective of the Convention, at the centre of the discourse surrounding the Ban Amendment was instrumental in breaking the long-standing political deadlock over the interpretation of Article 17 paragraph 5 of the Convention.

Immediately following the meeting, Switzerland and Indonesia launched a 'Country-Led Initiative to Improve the Effectiveness of the Basel Convention' (CLI), and invited key players to discuss and develop recommendations for consideration by the Tenth Meeting of the Conference of the Parties 'for a way forward to ensure that the transboundary

---

[33]   See <http://archive.basel.int/meetings/cop/cop7/commVII2/index.html> (last accessed on 11 August 2015).

[34]   See the Report of the Ninth Meeting of the Conference of the Parties, paragraph 77, available at <http://archive.basel.int/meetings/cop/cop9/docs/39e-rep. pdf> (last accessed on 11 August 2015).

[35]   Annex to Decision IX/26, the Report of the Ninth Meeting of the Conference of the Parties, available at <http://archive.basel.int/meetings/cop/cop9/docs/39e-rep.pdf> (last accessed on 11 August 2015).

movements of hazardous wastes, especially to developing countries and countries with economies in transition, constitute an environmentally sound management of hazardous wastes, as required by the Basel Convention'.[36] Through a succession of three informal meetings, experts from all five United Nations regional groups analyzed the reasons for the transboundary movements of hazardous wastes where ESM cannot be ensured, and elaborated several options on a way forward. By the time of the Tenth Meeting of the Conference of the Parties, a draft omnibus decision was before the parties, comprising seven sections: the entry into force of the Ban Amendment, including a proposed interpretation of paragraph 5 of Article 17; the development of standards and guidelines for ESM; the provision of further legal clarity; the further strengthening of the Basel Convention regional and coordinating centres; combating illegal traffic; assisting vulnerable countries; and capacity-building.[37] By the time of the opening of the Tenth Meeting of the Conference of the Parties, everyone's attention was focused on the possibility to, at last, overcome the challenges associated with the entry into force of the Ban Amendment and to reach agreement on a broader set of steps that would strengthen the achievement of ESM.

A second building block towards the historical outcomes of the Tenth Meeting of the Conference of the Parties was the launch, during the Ninth Meeting of the Conference of the Parties, of negotiations on a Strategic Framework for the implementation of the Basel Convention for 2012–21.[38] These negotiations were to provide parties with the opportunity to evaluate the effectiveness of the Convention and to define a new strategic direction for the Convention in the light of the evolving needs of the parties to the Convention, as well as the changing scientific, environmental, technical and economic circumstances under which the Convention was working.

Finally, supplementing the efforts under the CLI towards the awaited breakthrough on the issue of the Ban Amendment and complementing the negotiations of the Strategic Framework, Colombia, in its capacity as host, placed the Tenth Meeting of the Conference of the Parties under the theme 'Prevention, minimization and recovery of wastes.' The focus on prevention and minimization of wastes was to put at the centre of

---

[36] See the Report of the First Meeting of the CLI to the Expanded Bureau, available at <www.basel.int/Implementation/LegalMatters/CountryLedInitiative/Meetings/tabid/2680/Default.aspx> (last accessed on 11 August 2015).

[37] See document UNEP/CHW.10/5, available at <http://archive.basel.int/meetings/cop/cop10/documents/05e.pdf> (last accessed on 11 August 2015).

[38] Decision IX/3.

attention of one pillar of the Convention that had until then received so little attention, that of the generation of wastes. With the addition of the issue of the recovery of wastes, parties were able to go beyond the unchallenged health and environmental benefits of the Basel Convention and explore the potential economic value associated with the ESM of wastes, such as turning wastes into valuable resources for future consumption or production; conserving scarce and valuable materials such as rare earth metals; and creating green jobs. This opportunity could only be seized if the international community was ready to embrace the economic potential associated with the environmentally sound recovery of wastes.

4.6   THE TENTH MEETING OF THE CONFERENCE OF THE PARTIES: THE HISTORICAL AGREEMENTS

Convened ahead of the Rio+20 Summit, the Tenth Meeting of the Conference of the Parties not only held high expectations with respect to its possible contribution to this event, it also held the promise of the adoption of a cluster of decisions that could have a decisive impact on the future of the Convention.[39]

Through the adoption of the CLI omnibus decision (Decision BC-10/3), the Conference of the Parties managed to reach consensus on the interpretation of paragraph 5 of Article 17 of the Convention, thereby opening the door to the entry into force of the Ban Amendment. For the amendment to enter into force, it must be ratified by at least 'three-fourths of those parties that were Parties at the time of the adoption of the amendment', namely three-quarters of those parties that were parties to the Convention at the time of the Third Meeting of the Conference of the Parties in 1995. In other terms, the ratification of the Ban Amendment by parties that were not parties to the Convention at that time may not count towards reaching the threshold for the entry into force of the amendment. At the time of writing this chapter, although 85 parties have ratified the amendment, an estimated ten additional ratifications by qualified parties are still necessary in order to reach the entry into force threshold.

Through the adoption of the Strategic Framework (Decision BC-10/2), the meeting underlined the contribution of the Convention to promoting

---

[39]   See for instance the remarks of the Executive Director of UNEP, available at <www.basel.int/COP10/tabid/1571/Default.aspx> (last accessed on 11 August 2015).

sustainable livelihoods and attaining the Millennium Development Goals. It further endorsed several guiding principles for the implementation of the Convention over the next decade, including the waste management hierarchy (prevention, minimization, reuse, recycling, other recovery including energy recovery, and final disposal), and waste management policy tools such as the recognition of wastes as a resource.

Finally, the meeting adopted the Cartagena Declaration on the Prevention, Minimization and Recovery of Hazardous Wastes and Other Wastes whereby parties, among other things, reaffirmed that:

> the safe and environmentally sound recovery of hazardous and other wastes that cannot as yet be avoided, represents an opportunity for the generation of employment, economic growth and the reduction of poverty insofar as it is done in accordance with the Basel Convention requirements, guidelines and decisions and will not create a disincentive for their prevention and minimization

and acknowledged that 'prevention, minimization and recovery of wastes advance the three pillars of sustainable development, and that fulfilment of the Basel Convention's objectives is an important contribution to the United Nations Conference on Sustainable Development in Rio de Janeiro in 2012'.[40]

It is safe to say that the meeting exceeded all expectations, not only because of the adoption of the CLI decision, of the Strategic Framework and of the Cartagena Declaration on the Prevention, Minimization and Recovery of Hazardous Wastes and Other Wastes, but because it ended early, which is no small feat in international settings even in context with fewer political controversies. The Tenth Meeting of the Convention and a new equilibrium between its environmental, social and economic objectives, an outcome that was welcomed by the entire international community gathered at the Rio+20 Summit.[41]

---

[40] See Annex IV to the Report of the Tenth Meeting of the Conference of the Parties, available at <http://archive.basel.int/meetings/cop/cop10/documents/28e.pdf> (last accessed on 11 August 2015).

[41] See United Nations, 'The Future We Want', Outcome document of the World Conference on Sustainable Development (Rio+20), General Assembly A/RES/66/288, (New York, 2012) para. 219, available at <http://rio20.net/wp-content/uploads/2012/06/N1238164.pdf> (last accessed on 11 August 2015).

5. Transboundary movements of wastes and end-of-life goods under WTO law

Mirina Grosz

EXECUTIVE SUMMARY

Re-use, recycling, as well as environmentally sound waste management and disposal operations have become important economic factors, particularly in industrialized countries. It is thus not surprising that an international market for waste materials has emerged; waste and end-of-life goods are regularly traded and shipped across borders for their disposal and recovery.

In addressing the transboundary movements of wastes and end-of-life goods from the viewpoint of the law of the World Trade Organization ('WTO') and the General Agreement on Tariffs and Trade ('GATT') in particular, this chapter first raises the issue that the notion of 'waste' has a relative connotation. What is perceived as worthless 'rubbish' by some may be a valuable and tradable commodity for others, and as such, wastes and end-of-life goods will generally fall within the broad scope of application of WTO law and the GATT. As a consequence, states imposing trade restrictions on the transboundary movements of wastes and end-of-life goods run the risk of breaching international trade law.

By first examining the compatibility of trade measures with general principles of the GATT, this chapter addresses questions that are bound to arise when applying concepts of the GATT to end-of-life materials. It then analyses the possibilities of and limitations to justifying trade-restrictive measures under Article XX GATT, according to which deviations from the GATT principles may be legitimate if a state can demonstrate that its measures are necessary to reach legitimate policy goals and are applied in a manner that does not constitute a means of arbitrary or unjustifiable discrimination or a disguised restriction on international trade. In doing so, this chapter raises questions on the role of the WTO Panels and the

Appellate Body in addressing uncertain risk situations that touch on environmental, social and ethical ('non-trade') concerns.

The chapter comes to the conclusion that while restrictions on cross-border movements of hazardous wastes and end-of-life goods are most likely to be justified when implemented with a view to environmental and human health concerns, justifying less clear-cut cases – for example, cases involving materials that are not commonly known as 'hazardous' or trade restrictions grounded primarily on ethical considerations – is a more ambitious task. This outcome is also in line with the legal grey areas of the regulatory frameworks on transboundary movements of wastes on an international and regional level, which do not regulate or control non-hazardous, 'green-listed' wastes to a wide extent.

5.1 WASTES AND END-OF-LIFE GOODS TRADED AS COMMODITIES?

The notion of 'waste' is used when addressing the valueless and useless discarded by-products of our everyday lives, the leftovers of both production and consumption, often characterized as dirty, smelly or unhygienic. Now that scientific and technological advancements have significantly improved the possibilities for waste management, however, waste materials are increasingly seen as sources of valuable raw materials.[1] As such they constitute the establishing pillars of an industry that is based on the extraction of resources from end-of-life materials and their re-use and recycling.[2] The concept of 'end-of-life goods' is more precise than the

---

[1]   Turning waste into a resource and improving the economy's circularity are important waste policy objectives in the European Union (EU). See, e.g., the recently adopted EU Action Plan for the Circular Economy which also includes legislative proposals (European Commission, 'Closing the loop – An EU action plan for the Circular Economy', COM[2015]614 final) as well as the Seventh Environment Action Programme (Decision No. 1386/2013/EU of the European Parliament and of the Council of 20 November 2013 on a General Union Environment Action Programme to 2020, 'Living well, within the limits of our planet'); see also European Commission, 'Roadmap to a Resource Efficient Europe', COM(2011)571 final; European Commission, 'The Raw Materials Initiative – Meeting our Critical Needs for Growth and Jobs in Europe', COM(2008)699 final; European Commission, 'Taking sustainable use of resources forward – A Thematic Strategy on the prevention and recycling of waste', COM(2005)666 final.

[2]   See, e.g., the EU's Raw Materials Initiative (n 1); see also European Commission, 'Tackling the Challenges in Commodity Markets and on Raw Materials', COM(2011)25 final, 18–19; see also Martin O'Brien, *A Crisis of*

notion of 'waste' in this respect, as it implies that these materials have merely reached the end of their days and were valued as 'goods' in a previous stage of their life cycle.

Grounded on the understanding that 'waste' has a relative connotation and that what is 'waste' or an 'end-of-life-good' in one part of the world might be a valuable product with a 'new life' somewhere else,[3] waste materials have become the drivers for sectoral branches and international markets, which draw on and make use of these valuation differences. Waste materials and end-of-life goods are regularly shipped across national borders and are traded on corresponding markets.[4] Indeed, despite the substantial risks that the transboundary movement of wastes can entail, trading such materials is often perceived as a means to efficiently allocate them to specialized disposal and recycling sites, thereby enabling their environmentally sound management ('ESM') and recovery on a global level. It is therefore not surprising that according to recent studies, transboundary movements of wastes and end-of-life goods have increased significantly. In the European Union ('EU') alone, exports of all notified waste (hazardous and non-hazardous) have more than doubled from 6.3 million tons in 2001 to 15.4 million tons in 2013.[5] The European Environment Agency estimates that in the period between 1999 and 2011, non-hazardous waste plastic exports from Member States of the EU grew by a factor of five, waste precious metal exports trebled, while waste iron and steel, copper, aluminium and nickel exports doubled.[6] The amount of exports of hazardous waste

---

Waste? Understanding the Rubbish Society (Routledge 2008) 70, 74; Pierre-Marie Dupuy and Jorge E. Viñuales, *International Environmental Law* (Cambridge University Press 2015) 221.

[3] Perceptions of dirt and pollution have indeed been understood as cultural categories and issues of class rather than physical realities. See Mary Douglas, *Purity and Danger: An Analysis of the Concepts of Pollution and Taboo* (Routledge 2002, reprinted 2008); see also Susan Strasser, *Waste and Want, A Social History of Trash* (Metropolitan Books 1999) 8–12, 136–40; O'Brien (n 2) 125–43; for an overview see Mirina Grosz, *Sustainable Waste Trade under WTO Law* (Brill/Nijhoff 2011) 7 ff.

[4] On the different drivers of the transboundary movements of waste see, e.g., ETC/SCP Working Paper 2/2012, 'Transboundary Shipments of Waste in the European Union' (November 2012) 27 ff and 36 ff; EEA Report No. 7/2012, 'Movements of Waste across the EU's Internal and External Borders' (6 November 2012) 20 ff; Grosz (n 3) 107 ff.

[5] See Waste Shipment Statistics, available at <http://ec.europa.eu/eurostat/statistics-explained/index.php/Waste_shipment_statistics> (last accessed on 30 July 2016).

[6] See EEA Report (n 4) 20–21.

from EU Member States to other EU Member States or out of the EU have increased by 86 per cent from 3.2 million tons in 2001 to 5.9 million tons in 2013.[7]

The relativity of the concept of 'waste' is acknowledged in the various international legal frameworks governing such materials' cross-border movements. They all apply a broad definition of 'waste' that may include 'valuable' substances or objects that can be the subjects of re-use and recovery operations and that can also be traded for this purpose.[8] In other words, the perception of wastes as materials and substances that can have an economic value and can be traded as commodities is generally accepted on an international level.[9] This raises the question whether wastes and end-of-life goods also fall within the scope of WTO law and the GATT[10] in particular, as will be addressed in the following section.

---

[7] In 2007, waste shipments peaked at 8 million tons (see Waste Shipment Statistics [n 5] Table 1).

[8] See, e.g., Basel Convention on the Control of Transboundary Movements of Hazardous Wastes and their Disposal, 22 March 1989, UNEP/IG.80/3, 28 ILM 657 (1989); OECD Decision of the Council concerning the Control of Transboundary Movements of Wastes destined for Recovery Operations, 19 March 2001, C(2001)107/FINAL; see also Regulation of the European Parliament and of the Council (EC) 1013/2006 on Shipments of Waste (14 June 2006) OJ L 190/1, 12.7.2006, p. 1 (hereinafter: 'EU Waste Shipment Regulation'). For an overview of these regulatory frameworks see Grosz (n 3) 20 ff, 136 ff and 422 ff.

[9] Ideas to conceptually preclude wastes from the scope of trade agreements in order to minimize their transboundary movements from the outset have not gained general recognition. See Katharina Kummer, *Transboundary Movements of Hazardous Wastes at the Interface of Environment and Trade* (UNEP 1994) 72; Jonathan Krueger, *International Trade and the Basel Convention* (Earthscan Publications 1999) 67–8.

[10] This chapter generally refers to 'GATT', meaning both the GATT 1994 (15 April 1994, Part of the Marrakesh Agreement Establishing the World Trade Organization, Annex 1A, 33 ILM 1153 [1994]) as well as the GATT 1947 (1 January 1948, UNTS 194). The provisions of the GATT 1947 have been incorporated into the GATT 1994 and continue to have legal effect as part of the GATT 1994.

5.2 WASTES AND END-OF-LIFE GOODS UNDER WTO LAW

5.2.1 Applicability of the GATT in Particular

The GATT covers the international trade in goods. However, the notion of 'goods' is not defined in WTO law.[11] Indications as to whether a specific commodity falls under the GATT can be derived from the Harmonized Commodity Description and Coding System, established by the World Customs Organization and referred to as the 'Harmonized System' ('HS').[12] The HS comprises approximately 5,000 commodity groups, each identified by a six-digit code, and classifies about 98 per cent of the merchandise in international trade.[13]

While 'waste materials' or 'end-of-life goods' do not constitute a separate HS category, waste materials are referred to as particular subcategories of specific goods. For example, reference is made to residues and waste from the food industries (HS chapter 23), or to recovered waste and scrap paper or paperboard (HS Code 4707) as well as waste, parings and scrap, of plastics (HS Code 3915). These classifications imply that waste and end-of-life goods will usually not be traded as such. However, once specified as tradable 'scrap papers', 'used tyres', 'metal parts', etc., the rules of the GATT will generally be applicable.[14] In other words, the phase of a good's life cycle is not decisive with respect to the question of whether the international trade rules of the WTO apply.

This was confirmed by the WTO dispute settlement bodies in two quite recent cases. In the *Brazil – Tyres* case,[15] retreaded and used tyres – i.e.

---

[11]   See also James Munro, 'Pushing the Boundaries of "Products" and "Goods" under GATT 1994: An Analysis of the Coverage of New and Unorthodox Articles of Commerce' (2013) 47 *Journal of World Trade* 1323; Grosz (n 3) 254–7.

[12]   See HS Nomenclature 2012 and 2017 Edition, available at <www.wcoomd. org> (last accessed on 30 July 2016).

[13]   See <www.wcoomd.org/en/topics/nomenclature/overview/what-is-the-harm onized-system.aspx> (last accessed on 30 July 2016).

[14]   Notably, transfers of end-of-life materials across national borders could also be perceived as part of a cross-border service supply falling within the scope of the GATS (General Agreement on Trade in Services, Part of the Marrakesh Agreement Establishing the World Trade Organization, Annex 1B, 15 April 1994). See Grosz (n 3) in particular at 261 ff and at 415 ff.

[15]   See WTO, *Brazil – Measures Affecting Imports of Retreaded Tyres – Report of the Panel* (12 June 2007) WT/DS332/R, as modified by WTO, *Brazil – Measures Affecting Imports of Retreaded Tyres – Report of the Appellate Body* (3 December 2007) WT/DS332/AB/R, both available at <http://docsonline.wto.org>.

'waste tyres' – were the subjects of controversy.[16] Neither the Panel nor the Appellate Body found it necessary to address the issue whether such materials fall within the scope of WTO law. They did not hesitate to apply provisions of the GATT to the case. Similarly, in the *China – Raw Materials* case the Panel assessed the question whether China's export duties and export quotas on 'scrap' products (magnesium scrap, manganese scrap, and zinc scrap) were justified pursuant to Article XX(b) GATT without further ado.[17]

Nevertheless, cases are conceivable in which the qualification of transboundary movements of waste materials as 'international trade' may seem questionable. This could occur in situations where the waste materials are not subject to a 'commercial transaction' in terms of a sale contract, but are rather transferred to another state as an 'environmental burden'. If waste materials are shipped abroad in order to ensure their correct treatment and disposal in a specialized facility, it could be argued that the materials are not transferred as valuable goods and are therefore not actually 'traded' in terms of WTO law.[18] However, it would be a rather challenging task to establish when a transboundary shipment does not constitute 'trade' and, as a consequence, does not fall within the field of application of trade law. Therefore, such constellations will presumably be limited to exceptional cases.[19]

5.2.2 Addressing Waste Trade Restricting Measures under WTO Law Principles

Governmental policy responses tackling the trade with wastes and end-of-life materials may vary and range from import restrictions and export restrictions to technical regulations and requirements that the traded commodities have to comply with, such as mandatory recycling schemes

---

[16] 'Retreaded tyres' are tyres that have been recycled. They are produced by stripping the worn tread from a used tyre's casing and replacing it with new material (see *Brazil – Tyres*, Panel Report [n 15] para. 2.1).

[17] See WTO, *China – Measures Related to the Exportation of Various Raw Materials – Report of the Panel* (5 July 2011) WT/DS394/R, WT/DS395/R, WT/DS398/R, paras 7.470 ff.

[18] See also the witness testimony of Robert Howse relating to NAFTA in the hearing before the Subcommittee on Environment and Hazardous Materials of the Committee on Energy and Commerce, US House of Representatives, regarding 'Three Bills Pertaining to the Transport of Solid Waste: H.R. 382, H.R. 411 and H.R. 1730' (23 July 2003) available at <www.gpo.gov/fdsys/pkg/CHRG-08hhrg89003/html/CHRG-108hhrg89003.htm> (last accessed on 30 July 2016).

[19] Grosz (n 3) 256–7.

to name just one example.[20] The following outline will particularly focus on import and export restricting measures and will address their compatibility with general principles of the GATT.

5.2.2.1 Import restrictions

Besides introducing straightforward import bans or import quotas for specific wastes, a state may also decide to impose price-based measures such as (environmental) taxes on imported goods.[21] Import license requirements as well as the imposition of fines for carrying out specific imports, or for transporting or storing 'prohibited' goods, can also impede free access to domestic markets.

A state is likely to stipulate a ban on specific imports if there are indications of particular risks they imply. Furthermore, a state could limit the import of wastes with the argument that its waste management facilities are not in a position to cope with such materials in an environmentally sound manner.[22]

The right of states to unilaterally prohibit imports of waste to protect their territory from hazardous substances has been interpreted as a general principle of customary international law.[23] It is also recognized by multilateral agreements such as the Basel Convention[24] and the

---

[20] For example, the German 'Verpackungsverordnung' ('Packaging Ordinance'), originally dated 12 June 1991 (Bundesgesetzblatt [BGBl.] I Nr. 36 S 1234; translated in 21 ILM 1135 [1992]), regulates the packaging of products and sets mandatory recycling requirements for packaging waste. Under this Ordinance, manufacturers of products are required to take back packaging wastes and to arrange for their recycling in a private waste collection system. Participating manufacturers mark their products with the well established 'green dot'. On the trade implications that such 'life cycle laws' may have see Mitsuo Matsushita, Thomas J. Schoenbaum, Petros C. Mavroidis and Michael Hahn, *The World Trade Organization, Law, Practice and Policy* (OUP, 3rd ed. 2015) 752–3; see also Grosz (n 3) 400 ff with an assessment under the TBT Agreement (Agreement on Technical Barriers to Trade, Part of the Marrakesh Agreement Establishing the World Trade Organization, Annex 1A, 15 April 1994) of technical regulations and standards as well as labelling schemes possibly applicable in the context of cross-border movements of wastes.

[21] See, e.g., Patricia Birnie, Alan Boyle and Catherine Redgwell, *International Law and the Environment* (OUP, 3rd ed. 2009) 796–801; see also Andrew Green and Tracey Epps, 'The WTO, Science, and the Environment: Moving towards Consistency' (2007) 10 *Journal of International Economic Law* 285, 290–99; see also Matsushita, Schoenbaum, Mavroidis and Hahn (n 20) 759 ff on environmental taxes.

[22] See, e.g., Art. 4(2)(g) of the Basel Convention. On import restrictions imposed on waste materials see Grosz (n 3) 363 ff and 381 ff.

[23] See, e.g., Grosz (n 3) 366 with further references.

[24] See the preambular para. 6 and Art. 4(1) of the Basel Convention.

EU Waste Shipment Regulation.[25] According to Article 4(5) of the Basel Convention in particular, Convention parties 'shall not permit hazardous wastes or other wastes to be exported to a non-[p]arty or to be imported from a non-[p]arty'. This provision was established to prevent party states from engaging in hazardous waste trading with states that, as non-parties, do not adhere to the provisions of the Basel Convention.[26] Similarly, imports of waste for disposal and recovery from so-called 'third countries' are generally prohibited by the EU Waste Shipment Regulation: waste is only accepted from countries that are parties to the Basel Convention and/or members of the Organisation for Economic Co-operation and Development ('OECD') and are thus bound by the OECD Council Decision C(2001)107/FINAL.[27] Furthermore, and in particular, an agreement between the trading parties is required to ensure the ESM of the objects and substances in question.[28]

Irrespective of the motivations underlying the trade impediments, import restrictions will often amount to breaches of the general prohibition of quantitative restrictions according to Article XI:1 GATT.[29] Furthermore, in cases where trade-restrictive measures differentiate between materials

---

[25]   See recital 9 of the EU Waste Shipment Regulation (n 8).

[26]   Katharina Kummer, *International Management of Hazardous Wastes, the Basel Convention and Related Legal Rules* (OUP 1995, reprinted 1999) 61–3; see also Tobias Bender, *Domestically Prohibited Goods, WTO-Rechtliche Handlungsspielräume bei der Regulierung des Handels mit im Exportland verbotenen Gütern zum Umwelt- und Verbraucherschutz* (Duncker & Humblot 2006) 399–401; David Wirth, 'Trade Implications of the Basel Convention Amendment Banning North-South Trade in Hazardous Wastes' (1998) *Review of European, Comparative and International Environmental Law* 237, 241–2.

[27]   See (n 8) above.

[28]   See also Grosz (n 3) 364 ff.

[29]   Art. XI:I GATT reads:

No prohibitions or restrictions other than duties, taxes or other charges, whether made effective through quotas, import or export licences or other measures, shall be instituted or maintained by any contracting party on the importation of any product of the territory of any other contracting party or on the exportation or sale for export of any product destined for the territory of any other contracting party.

On quantitative restrictions see, e.g., Matsushita, Schoenbaum, Mavroidis and Hahn (n 20) 239 ff; Petros C. Mavroidis and Mark Wu, *The Law of the World Trade Organization (WTO), Documents, Cases and Analysis* (West Academic Publishing, 2nd ed. 2013) 59 ff; Raj Bhala, *International Trade Law: An Interdisciplinary Non-Western Textbook, Volume One* (LexisNexis 2015) 723 ff.

stemming from different countries, both the most-favoured-nation principle ('MFN principle') pursuant to Article I GATT[30] and the national treatment principle according to Article III GATT[31] may be at issue.[32] As key provisions of the multilateral trading system, they prohibit the different treatment of 'like products' with diverging origins and are grounded in the idea of equality, equal treatment and non-discrimination.[33]

The concept of 'likeness' is a fundamental element of both the MFN and the national treatment principles. It is due to commodities' likeness that they become comparable and that discriminatory measures can be assessed.[34] If waste materials are traded as commodities, interesting

---

[30]   Art. I:1 GATT reads:

With respect to customs duties and charges of any kind imposed on or in connection with importation or exportation or imposed on the international transfer of payments for imports or exports, and with respect to the method of levying such duties and charges, and with respect to all rules and formalities in connection with importation and exportation, and with respect to all matters referred to in paragraphs 2 and 4 of Article III, *any advantage, favour, privilege or immunity granted by any contracting party to any product originating in or destined for any other country shall be accorded immediately and unconditionally to the like product originating in or destined for the territories of all other contracting parties* (emphasis added.)

[31]   Art. III:2 GATT reads:

The products of the territory of any contracting party imported into the territory of any other contracting party shall not be subject, directly or indirectly, to internal taxes or other internal charges of any kind in excess of those applied, directly or indirectly, to like domestic products (. . .)

Art. III:4 reads:

The products of the territory of any contracting party imported into the territory of any other contracting party shall be accorded treatment no less favourable than that accorded to like products of national origin in respect of all laws, regulations and requirements affecting their internal sale, offering for sale, purchase, transportation, distribution or use . . .

[32]   See also Grosz (n 3) 363 ff.
[33]   See Matsushita, Schoenbaum, Mavroidis and Hahn (n 20) 155–6; see also Mavroidis and Wu (n 29) 121 ff and 215 ff; Bhala (n 29) 671.
[34]   On the concept of 'likeness' see the Report of the Working Party on Border Tax Adjustments of 2 December 1970, BISD 18S/97, GATT Doc L/3463, para. 18; see also, e.g., William J. Davey and Joost Pauwelyn, 'MFN-Unconditionality: A Legal Analysis of the Concept in View of its Evolution in the GATT/WTO Jurisprudence with Particular Reference to the Issue of "Like Product"' in Thomas Cottier and Petros C. Mavroidis (eds), *Regulatory Barriers and the Principle of Non-Discrimination in World Trade Law: Past, Present, and Future* (The University of Michigan Press 2000) 25–36; Bhala (n 29) 503 ff.

questions are bound to arise in this regard.[35] For example, when comparing the markets of new products, re-usable goods and wastes, the issue could be raised whether re-usables should be treated like new goods, or whether their prior life cycle has altered their physical characteristics in such a manner that they should be traded on another market. Can wastes and non-wastes be treated as like products, or should the 'likeness test' be limited to objects and substances in the same 'phase of life'?[36] Furthermore, is not the comparison of physical characteristics of tradable goods too narrow a focus to assess the possible risks of their cross-border movements?[37] Indeed, the risks generated by waste often stem from the materials' *treatment* as waste, not necessarily by the materials' physical characteristics. For example, unsafe and environmentally unsound storage, transport and management can cause certain hazards. Additionally, disposal and recovery operations can also entail risks.[38] These are just some of the complex issues that the cross-border movements of waste materials may raise under WTO law – questions that ultimately have to be addressed on a case-by-case basis.

5.2.2.2 Export restrictions

A variety of reasons exists for restricting the export of wastes or end-of-life goods.[39] A state may opt for such trade measures based on concerns that the environmentally safe and sound management of the materials in question would not be guaranteed at the location of destination. Furthermore, export restraints could be implemented with the expectation that such trade measures would influence the importing states' waste management operation standards.[40] It is also conceivable that states decide to prohibit the exports of materials based on moral, environmental or human health considerations, or something like a 'feeling of responsibility' for disposing of the materials that have been generated on their territory.[41]

Noteworthy reasons for export restrictions were also at issue in the

---

[35] See also Grosz (n 3) 389 ff on the implications that the waste trade may have for the concepts of likeness and process and production methods.

[36] Ibid., 391.

[37] See also Robert Howse and Donald Regan, 'The Product/Process Distinction – An Illusory Basis for Disciplining "Unilateralism" in Trade Policy' (2000) 11 *European Journal of International Law* 249, 260.

[38] See also Grosz (n 3) 393 with further references.

[39] On export restrictions affecting the cross-border movements of waste see Grosz (n 3) 371 ff and 383 ff.

[40] See also Birnie, Boyle and Redgwell (n 21) 788; Matsushita, Schoenbaum, Mavroidis and Hahn (n 20) 536–7.

[41] See also Grosz (n 3) 372–3 and 376–8.

China – Raw Materials case. According to China, the primary production of magnesium, manganese and zinc is highly polluting, energy intensive and causes significant health risks.[42] In contrast, the metals' production using recycled scrap 'is significantly less polluting and more energy efficient' and reduces the risks related to the use of crude ores.[43] China therefore argued that its export restraints on scrap products were necessary, both to promote its recycling industry by ensuring a steady supply of scrap products as well as to reduce pollution.[44]

Export restrictions imposed by a WTO Member State may breach Article XI:1 GATT[45] when made effective through export bans or export quotas, export licenses and other measures which limit the exportation or sale for export of any product destined for the territory of another WTO contracting party.[46] According to the wording of Article III GATT,[47] however, the national treatment principle only applies to imported goods. By contrast, it is accepted in legal doctrine and practice that no apparent reason exists to also apply this limitation to the MFN principle.[48] As a consequence, if customs duties and charges as well as rules and formalities imposed on or in connection with exports of wastes differentiate between the different countries of destination, such measures may be in breach of Article I:1 GATT.[49]

The Basel Convention aims at minimizing cross-border movements of hazardous wastes by also restricting their exports. Article 4(5), for example, stipulates that parties are not permitted to export hazardous wastes to a non-party. Furthermore, the so-called 'Ban Amendment' prohibits transboundary movements of hazardous wastes between countries listed in Annex VII of the Basel Convention and those countries not listed in Annex VII.[50] Annex VII includes OECD Member States, Member States of the EU as well

---

[42]   *China – Raw Materials*, Panel Report (n 17) paras 7.470–7.471, 7.494 and 7.592.

[43]   Ibid., paras 7.471 and 7.592.

[44]   Ibid., paras 7.470–7.472 and paras 7.478 ff with the Panel's assessment.

[45]   See (n 29) above.

[46]   For an overview of WTO case law on export restrictions see, e.g., Baris Karapinar, 'Defining the Legal Boundaries of Export Restrictions: A Case Law Analysis' (2012) 15 *Journal of International Economic Law* 443; see also Matsushita, Schoenbaum, Mavroidis and Hahn (n 20) 537–40.

[47]   See (n 31) above.

[48]   See *Japan – Trade in Semi-Conductors* (1988) GATT BISD 35S/116; Davey and Pauwelyn (n 34) 17; see also Matsushita, Schoenbaum, Mavroidis and Hahn (n 20) 542 regarding export tariffs.

[49]   See (n 30) above.

[50]   On export restrictions according to the Basel Convention and on the Ban Amendment in particular see Grosz (n 3) 371 ff and 384 ff with further references.

as Liechtenstein. It is expected that Annex VII countries are in a position to ensure environmentally sound waste management operations.[51]

The Ban Amendment has not yet entered into force.[52] It is however widely respected and has been incorporated into the EU Waste Shipment Regulation, which establishes different rules for the waste trade within the EU and the trade with third countries, thereby additionally distinguishing between waste transfers destined for disposal or recovery operations.[53]

From a WTO law vantage point, such Basel Convention inspired export prohibitions are in breach of Article XI GATT.[54] Furthermore, distinguishing between particular country categories without assessing the actual situation in these countries on a case-by-case basis – an approach adopted by both the Ban Amendment of the Basel Convention as well as the EU Waste Shipment Regulation – may result in discrimination against countries where the same conditions prevail.[55]

5.2.3　Justifying Deviations from GATT Principles

The multilateral trading system as it presents itself today acknowledges deviations from its own principles if certain important policy goals are given priority, provided that the measures enacted meet the applicable legal requirements.[56] Breaches of GATT principles can be justified under Article XX GATT in particular.

According to the Panel's and the Appellate Body's case law, the assessment of a trade measure under Article XX GATT follows a tiered analysis which can be broken down into three steps: first, a breach of general GATT principles needs to be based on one of the motives and conditions for restricting trade as listed in the paragraphs (a)–(j) of Article XX GATT. Second, the measure in question is required to correlate with the

---

[51]　For further details see Grosz (n 3) 384–5.

[52]　See <www.basel.int/Countries/StatusofRatifications/BanAmendment/tabid/ 1344/Default.aspx> (last accessed on 30 July 2016); for an overview on the ongoing controversy as to the Amendment's entry into force see Ulrich Beyerlin and Thilo Marauhn, *International Environmental Law* (Hart Publishing/Verlag CH Beck 2011) 217–18.

[53]　For further details see Grosz (n 3) 385 with further references.

[54]　Ibid. (n 3) 372 with further references.

[55]　Kummer (n 26) xxx–xxxiii; Krueger (n 9) 71–2; Shawkat Alam, *Sustainable Development and Free Trade, Institutional Approaches* (Routledge 2008) 200; Grosz (n 3) 385.

[56]　Thomas Cottier and Matthias Oesch, *International Trade Regulation, Law and Policy in the WTO, the European Union and Switzerland* (Staempfli 2005) 347, 428; see also Mavroidis and Wu (n 29) 285–6.

subject of protection. According to the legal text of Article XX GATT, such a relationship is acknowledged if the measure is 'necessary' to achieve the policy goal[57] or 'related to' the pursuit of the regulatory objectives.[58] Third, exceptions to the GATT principles ultimately also have to be in compliance with the requirements of the introductory clause of Article XX GATT, commonly termed the 'chapeau'.[59]

5.2.3.1 Legitimate policy goals

Restrictions on transboundary movements of waste and end-of-life goods are likely to be based on Article XX(b) GATT. This provision lists 'human, animal or plant life or health' as policy goals that can justify deviations from the GATT principles.

The scope of Article XX(b) GATT has been interpreted rather broadly.[60] In the *Brazil – Tyres* case for example, Brazil argued that its import ban was justified as a necessary measure for the protection of 'human life and health and the environment',[61] because it would reduce waste tyre volumes and, by so doing, also decrease associated risks such as the incidence of mosquito-borne diseases as well as fire hazards and environmental contamination.[62] The Panel accepted the reference made to the 'environment', despite the fact that the environment as such is not mentioned in Article XX(b) GATT. However, it continued by stating that Brazil had to substantiate 'the existence not just of risks to "the environment" generally, but specifically of risks to animal or plant life or health'.[63] This interpretation of Article XX(b) GATT was subsequently also accepted by the Appellate Body.[64]

Article XX(g) GATT additionally justifies breaches of GATT principles caused by measures 'relating to the conservation of exhaustible natural resources if such measures are made effective in conjunction with

---

[57] See paras (a), (b) and (d) of Article XX GATT.
[58] See paras (c), (g) and (e), ibid.
[59] See also Grosz (n 3) 430 ff with further references.
[60] See, e.g., Catherine Button, *The Power to Protect, Trade, Health and Uncertainty in the WTO* (Hart Publishing 2004) 24–40; Jochem Wiers, *Trade and Environment in the EC and the WTO, A Legal Analysis* (Europa Law Publishing 2003) 184–6; Grosz (n 3) 437; see also Michael J. Trebilcock, Robert Howse and Antonia Eliason, *The Regulation of International Trade* (Routledge, 4th ed. 2013) 664 ff on Article XX GATT and environmental concerns.
[61] See in particular *Brazil – Tyres*, Panel Report (n 15) para. 7.44.
[62] Ibid., paras 4.11 ff.
[63] Ibid., para. 7.45.
[64] See, e.g., *Brazil – Tyres*, Appellate Body Report (n 15) paras 140, 151, 171, 179, and 210.

restrictions on domestic production or consumption'. If 'waste materials' were found to include 'natural resources' according to this provision, however, would not the possibility of their recovery speak against the resources' exhaustibility? As this question shows, in the context of the transboundary movement of waste materials the issue may arise whether the notion of 'exhaustible natural resources' should only apply to raw materials, i.e., resources that are 'freshly' extracted and that have not yet been processed, recovered or used. China in fact built on such a line of argument in its first written submission in the *China – Raw Materials* case.[65] Notably, however, in the *China – Rare Earths* case a few years later, the Panel acknowledged that a measure may 'relate to the conservation of' exhaustible natural resources even if that resource in its raw form is not the direct subject of the measure.[66] But even if waste materials could be defined as 'exhaustible natural resources', current tendencies to promote trade in re-used and recycled products (with the objective of protecting natural resources by enhancing the supply of alternatives to primary raw materials) could make attempts to justify restrictions on such products' import or export under Article XX(g) GATT an ambitious task.[67]

Imports and exports of waste can also have human rights implications.[68] Human rights are however not expressly covered by the wording of Article XX GATT.[69] Nevertheless, ethical concerns could prompt the question whether certain trade measures can be justified as 'necessary to protect public morals' under Article XX(a) GATT. Is it 'right' to accept wastes from other countries together with the potential risks that such transfers imply? And is it morally acceptable to export such materials to states that were not involved in the wastes' generation? The answers to such questions may differ from society to society and depending on the circumstances of the case.[70] However, international acceptance of the provisions of the Basel Convention, reiterated concerns regarding the protection of

---

[65] See the reference made to paras 101 and 107 of China's first written submission in footnotes 572 and 573 of *China – Raw Materials*, Panel Report (n 17).

[66] WTO, *China – Measures related to the Exportation of Rare Earths, Tungsten, and Molybdenum – Report of the Panel* (26 March 2014) WT/DS431/R, WT/DS432/R, WT/DS433/R, para. 7.247; see also the ruling in *US – Auto Taxes* according to which carbon fuels fell within the scope of Article XX(g) GATT, because carbon fuels are made from petroleum (WTO, *US – Taxes on Automobiles – Report of the Panel* [unadopted, circulated 11 October 1994] WT/DS31/R).

[67] Grosz (n 3) 438–9.

[68] Ibid., 209 ff.

[69] Article XX(e) GATT does however address products of prison labour. On human rights and Article XX GATT see, e.g., Bhala (n 29) 881 ff.

[70] WTO, *US – Measures Affecting the Cross-Border Supply of Gambling*

developing countries from accumulations of hazardous wastes, as well as public indignation caused by reports of illegal waste dumps reveal a common concern for such issues, particularly in cases where high risks are implied. It follows that legal interpretation of Article XX(a) GATT does not preclude a reading of 'public morals' as including 'beliefs' of the importing or exporting countries about the wrongfulness of trading particular materials and substances.[71] A different question is, of course, what the implications for the WTO dispute settlement mechanisms are if the WTO Panels and the Appellate Body decide on complex 'non-trade' issues from within the international trade law regime.[72]

5.2.3.2   Balancing legitimate policy goals under the exception provisions

Trade measures have to contribute to the realization of the legitimate policy goals pursued in order to be justified. According to Article XX(a) and Article XX(b) GATT, the provisions that are particularly addressed in this section, GATT-inconsistent measures can be justified if the adopted trade measures are 'necessary' to achieve the legitimate policy goals they envisage.

In order to assess the 'necessity' of a measure for a particular policy goal, the so-called 'three-step test' has been referred to, which encompasses a process of weighing and balancing different factors including: (i) the measure's contribution to the realization of the policy objective pursued; (ii) the importance of the interests and values protected by the measure; and (iii) the trade impact of the measure applied. This approach was set out by the Appellate Body in the *Korea – Beef* case[73] and was reiterated in several later rulings.[74] It is used to structure the following section that touches on certain aspects of the balancing test that are particularly interesting when

---

and Betting Services – Report of the Panel (10 November 2004) WT/DS285/R, para. 6.461 on Art. XIV GATS.

71   See also Robert Howse and Joanna Langille, 'Permitting Pluralism: The Seal Products Dispute and Why the WTO Should Accept Trade Restrictions Justified by Noninstrumental Moral Values' (2012) 37 *Yale Journal of International Law* 367 ff, particularly at 413–14, 427–32 on the *Seal Products* case (WTO, *EC – Measures Prohibiting the Importation and Marketing of Seal Products – Report of the Appellate Body* [22 May 2014] WT/DS400/AB/R, WT/DS401/AB/R); on the Article XX(a) GATT morality exception see also Grosz (n 3) 440–42; Bhala (n 29) 891 ff.

72   See, e.g., Grosz (n 3) 488 ff with further references.

73   WTO, *Korea – Measures Affecting Imports of Fresh, Chilled and Frozen Beef – Report of the Appellate Body* (11 December 2000) WT/DS169/AB/R, WT/DS161/AB/R.

74   See, e.g., *China – Raw Materials,* Panel Report (n 17) paras 7.481 ff with references to *Brazil – Tyres,* Appellate Body Report (n 15) para. 178.

assessing trade with waste and end-of-life goods under the GATT. The *Brazil – Tyres* case provides for an interesting case example in this context.

5.2.3.2.1   Measure's contribution to the realization of the policy objective pursued   In the *Brazil – Tyres* case, the WTO dispute settlement bodies were confronted with the question whether Brazil's import restrictions on retreaded tyres contributed to its regulatory goal of reducing the risks stemming from waste tyre accumulation 'to the maximum extent possible'.[75] Brazil's counterparty in this case, the European Communities ('EC'), argued that only incorrectly managed tyres, i.e., abandoned tyres or tyres negligently placed in monofills, could lead to the risks that Brazil claimed to address with its trade-restrictive measures.[76]

The Panel, however, took the view that it is a fact of reality 'that waste tyres get abandoned and accumulated and that risks associated with accumulated waste tyres', such as the spread of mosquito-borne diseases and tyre fires, 'exist' in countries with tropical climates such as Brazil.[77] The Panel also accepted that the import ban adopted was capable of contributing to the reduction of the overall number of waste tyres generated in Brazil, due to the fact that it targeted retreaded tyres which 'by definition' possess a shorter lifespan than new tyres.[78] Additionally, the measure was perceived as providing incentives for domestic retreaders to retread more domestic used tyres than imported tyres.[79] In sum, the Panel acknowledged that a reduction in the number of waste tyres would contribute to the protection of the environment and human health in Brazil.[80]

On appeal, the Appellate Body found the 'qualitative' analysis adopted by the Panel to be justified. It particularly did not require a quantification of the import ban's contribution to Brazil's policy objective.[81] It held that import bans sufficiently contribute to the achievement of the policy objectives under Article XX(b) GATT 'where there is a genuine relationship of ends and means between the objective pursued and the measure at issue'[82]

---

[75]   *Brazil – Tyres*, Panel Report (n 15) para. 7.108; *Brazil – Tyres*, Appellate Body Report (n 15) para. 144.

[76]   *Brazil – Tyres*, Panel Report (n 15) para. 7.63.

[77]   Ibid., para. 7.67; see also paras 7.61, 7.64, 7.71 and 7.80. The Panel, inter alia, based its findings on reports of the World Health Organization and on the Basel Convention Technical Guidelines on the Identification and Management of Used Tyres.

[78]   *Brazil – Tyres*, Panel Report (n 15) para. 7.130.

[79]   *Brazil – Tyres*, Panel Report (n 15) paras 7.115–7.142.

[80]   Ibid., paras 7.146–7.148.

[81]   *Brazil – Tyres*, Appellate Body Report (n 15) paras 145–47 and 152–55.

[82]   Ibid., para. 145.

and where the measure's contribution to the achievement of the objective is material.[83] In view of Brazil's comprehensive strategy to deal with waste tyres – of which the import ban appeared to be just one of the key elements – the Appellate Body held that:

> (. . .) in the short-term, it may prove difficult to isolate the contribution to public health or environmental objectives of one specific measure from those attributable to the other measures that are part of the same comprehensive policy. Moreover, the results obtained from certain actions – for instance, measures adopted in order to attenuate global warming and climate change, or certain preventive actions to reduce the incidence of diseases that may manifest themselves only after a certain period of time – can only be evaluated with the benefit of time.[84]

This statement was singled out as a first-time recognition of the right of WTO members to set ambitious environmental policy goals, even if their attainment may have trade-restrictive effects and even if their achievement cannot be quantified within a short time span.[85] The approach adopted by the Appellate Body was however also criticized (sometimes sharply) as leaving open essential methodological questions regarding the test to be applied when assessing a measure's contribution to the policy goal pursued.[86]

5.2.3.2.2 The importance of the interests protected According to the weighing and balancing test applied by the Appellate Body, 'the more vital or important common interests or values are, the easier it would be to accept as "necessary" a measure designed as an enforcement instrument'.[87] The

---

[83]   Ibid., para. 210. In the *China – Raw Materials* case, the Panel reiterated this approach (see *China – Raw Materials*, Panel Report [n 17] para. 7.518).

[84]   *Brazil – Tyres*, Appellate Body Report (n 15) para. 151.

[85]   Sébastien Thomas, 'Trade and the Environment under WTO Rules after the Appellate Body Report in Brazil – Retreaded Tyres' (2009*) Journal of International Commercial Law and Technology* 42–3, 45, 48–9; Markus W. Gehring, 'Sustainable Development in World Trade Law, A Short History' in Hans Christian Bugge and Christina Voigt (eds), *Sustainable Development in International and National Law* (Europa Law Publishing 2008) 289–90; Jefferey Atik, 'Trade and Health', in Daniel Bethlehem, Donald McRae, Rodney Neufeld and Isabelle van Damme (eds), *The Oxford Handbook of International Trade Law* (OUP 2009) 614–15.

[86]   See Chad P. Bown and Joel P. Trachtman, 'Brazil – Measures Affecting Imports of Retreaded Tyres: A Balancing Act' (2009) 8 *World Trade Review* 85, 125, 129–31. For an overview see also Grosz (n 3) 467–9.

[87]   *Korea – Beef*, Appellate Body Report (n 73) para. 162. See also WTO, *EC – Measures Affecting Asbestos and Asbestos-Containing Products – Report of the Appellate Body* (12 March 2001) WT/DS135/AB/R, para. 172; *Brazil – Tyres*,

assessment of applied regulatory goals' 'importance' is of course not without its problems.[88] In a nutshell, legal doctrine and WTO case law seem to suggest that if trade regulations are based on concerns for human health and the environment, and particularly if scientifically proven dangerous materials are at issue, the importance of the protected interests will generally be acknowledged. It would presumably be difficult to justify trade restrictions with the purpose of protecting 'public morals' alone. According to the same line of reasoning, it would also seem more ambitious to justify restrictions of the trade with non-hazardous recoverable resources under the WTO legal framework, than restrictions of hazardous wastes destined for specialized treatment and disposal.[89]

5.2.3.2.3 The trade impact of the measure   Justifying a trade measure with a restrictive effect as 'necessary' is generally perceived to be more intricate than justifying a measure that only has a slight impact on trade relations.[90] When examining the possible consequences of a particular trade instrument, legal doctrine and practice often assess whether a 'reasonably available' alternative exists that would achieve the same end and have a less trade-restrictive effect than the measure applied.[91] The *Brazil – Tyres* case once more provides for an interesting example.

In this case, the EC suggested several alternatives to the import prohibitions that Brazil implemented. The alternatives included measures to encourage the retreading of domestic passenger car tyres, measures to reduce the use of cars altogether (for example by promoting public transportation), measures aiming at a longer and safer use of retreaded tyres, as well as measures to improve the management of waste tyres (such as improved collecting and disposal systems, controlled landfilling, stockpiling, energy recovery and material recycling).[92] However, given Brazil's goal to reduce 'to the maximum extent possible'[93] the risks associated with waste tyre accumulation, the Panel found that no alternative measure would

---

Panel Report (n 15) paras 7.108–7.112 and para. 7.210; *Brazil – Tyres*, Appellate Body Report (n 15) paras 179 and 210; *China – Raw Materials*, Panel Report (n 17) paras 7.842–7.843.

[88]   See Grosz (n 3) 459–74, 469–70.
[89]   Ibid., 469 with further references and 442 ff.
[90]   *Korea – Beef*, Appellate Body Report (n 73) para. 163.
[91]   See, e.g., *EC – Asbestos*, Appellate Body Report (n 87) paras 169 ff; WTO, *US – Measures Affecting the Cross-Border Supply of Gambling and Betting Services – Report of the Appellate Body* (7 April 2005) WT/DS285/AB/R, para. 308.
[92]   *Brazil – Tyres,* Panel Report (n 15) paras 7.159–7.161.
[93]   Ibid., para. 7.108; *Brazil – Tyres*, Appellate Body Report (n 15) para. 144.

achieve the same outcome and be reasonably available.[94] Both the Panel and the Appellate Body particularly acknowledged that management or disposal operations would require substantial resources, technologies and know-how, would not lead to the reduction in the number of waste tyres generated by 'imported short-lifespan retreaded tyres' and would ultimately not avoid the risks stemming from imported retreaded tyres.[95] The Panel and the Appellate Body therefore found the suggested alternative measures to be appropriate as possible cumulative instead of substitutable measures.[96]

This outcome can be interpreted as acknowledging regulatory leeway for states to adopt comprehensive waste management policies, of which trade regulations may be just one possible element. Arguably, however, the way Brazil's regulatory goals were framed in this case also influenced the outcome of the necessity test applied; the ambitious formulation of Brazil's policy objectives made it particularly difficult to find WTO-consistent alternatives that would have provided the same level of protection.[97]

5.2.3.3 The Chapeau Test

Exceptions to the GATT principles have to be consistent with the chapeau of Article XX GATT. According to the chapeau's wording, a measure may not be 'applied in a manner which would constitute a means of arbitrary or unjustifiable discrimination between countries where the same conditions prevail, or a disguised restriction on international trade (. . .)'. Succinctly put, the chapeau reiterates the principle of non-discrimination. As an 'introductory remark' to the exception provisions, it allows tackling the possibility of abuse of the exceptions for protectionist trade measures[98] and can be interpreted as an expression of the principle of good faith.[99]

So far, a consistent test to examine the chapeau clause has not been devel-

---

[94]   *Brazil – Tyres*, Panel Report (n 15) paras 7.166, 7.172 and 7.212.

[95]   Ibid., paras 7.168 and 7.212. See also *Brazil – Tyres*, Appellate Body Report (n 15) paras 173–175.

[96]   *Brazil – Tyres,* Panel Report (n 15) paras 7.172 and 7.169; see also *Brazil – Tyres,* Appellate Body Report (n 15) para. 172.

[97]   See Grosz (n 3) 471–3. This reading of the *Brazil – Tyres* case also seems to be corroborated by the Panel's findings in the *China – Raw Materials* case, according to which China had not established that the available WTO-consistent alternatives could not provide the level of protection it had chosen to employ (*China – Raw Materials*, Panel Report [n 17] paras 7.564 ff).

[98]   See WTO, *US – Standards for Reformulated and Conventional Gasoline – Report of the Appellate Body* (29 April 1996) WT/DS2/AB/R, p. 22; see also WTO, *US – Import Prohibition of Certain Shrimp and Shrimp Products – Report of the Appellate Body* (12 October 1998) WT/DS58/AB/R, para. 156.

[99]   *US – Shrimp*, Appellate Body Report, ibid., para. 158.

oped. An interesting approach was applied by the Appellate Body in the *Brazil – Tyres* case[100] when it was confronted with the following situation: Brazil's import ban on retreaded tyres had not only been challenged in front of the WTO but also by Uruguay under the Mercado Común del Sur (Mercosur) agreement. The Mercosur tribunal found that the import ban constituted a prohibited trade restriction. Therefore, in order to comply with this ruling, Brazil exempted tyres from Mercosur Member States from the application of the import ban at issue. Additionally, Brazilian courts issued several injunctions which permitted the import of significant volumes of used tyres. It was the fact that Brazil allowed exceptions to the import restrictions that subsequently led to the Appellate Body's decision. The Appellate Body found that, because the Mercosur exemptions did not bear any relationship with the policy goals pursued under Article XX(b) GATT, they resulted in the import ban's arbitrary or unjustifiably discriminatory application.[101] According to the Appellate Body:

> (. . .) there is arbitrary or unjustifiable discrimination when a measure provisionally justified under a paragraph of Article XX is applied in a discriminatory manner 'between countries where the same conditions prevail', and when the reasons given for this discrimination bear no rational connection to the objective falling within the purview of a paragraph of Article XX, or would go against that objective.[102]

In other words, according to this ruling, if states can substantiate that the trade measures applied are used with the rationale of achieving the legitimate policy goals invoked, the measures are more likely to be regarded as consistent with the chapeau of Article XX GATT.

5.3 CONCLUSION

Hazardous and non-hazardous materials, used and second-hand products, wastes and natural resources can all be traded under the WTO legal framework when valued as commodities or as the subjects of waste management services. As this brief study has shown, the GATT regime does not prohibit states from tackling waste imports or exports with trade measures: states are not forced to import goods they perceive as dangerous

---

[100] See Grosz (n 3) 474 ff; see also Arwel Davies, 'Interpreting the Chapeau of GATT Article XX in Light of the "New" Approach in Brazil – Tyres' (2009) *Journal of World Trade* 507, 509.

[101] *Brazil – Tyres,* Appellate Body Report (n 15) paras 228–33 and 246–47.

[102] Ibid., para. 227.

or for which they do not have the infrastructure, the technologies and the know-how required. By the same token, states are not required to export materials they prefer to keep. Trade restrictions limiting the transboundary movements of commodities may breach GATT principles. However, if a state can demonstrate that its measures are necessary to reach legitimate policy goals and are applied in a manner that does not constitute a means of arbitrary or unjustifiable discrimination or a disguised restriction on international trade, such deviations from the GATT principles are justified.

The particular circumstances of a case will be decisive for its legal assessment. But in view of the existing case law and legal doctrine as well as the international regulatory frameworks in place, tendencies are discernible that restrictions on cross-border movements of hazardous wastes and end-of-life goods are most likely to be justified when implemented with a view to protecting human health and the environment. Such 'vital' concerns and unquestioned policy objectives may help legitimize trade measures as 'necessary'. Furthermore, if scientific evidence exists to substantiate the alleged risks and if the dangers associated with certain materials are recognized on an international level (for example under the Basel Convention), prospects are good that a state's discretion to restrict or even ban such imports or exports would be acknowledged. By contrast, justifying measures restricting trade with non-hazardous wastes and end-of-life goods tends to be a more ambitious task.

This difference in addressing 'hazardous' and 'non-hazardous' materials and substances under WTO law corresponds to the regulatory approaches adopted by the international legal frameworks that specifically address the cross-border movements of waste: shipments of raw materials and recyclable resources perceived as 'goods' are regulated more liberally to ensure their unhindered flow across national borders in order to promote strong and specialized waste management and treatment industries. By contrast, the transboundary movement of potentially hazardous materials (i.e., 'bads') is subject to more stringent regulations that focus particularly on the polluting and dangerous effects that such shipments may have.[103]

Where the distinction between 'hazardous' and 'non-hazardous' is more ambiguous, complex questions arise and the regulatory responses often

---

[103]   See also Grosz (n 3) 275–6 and 115 ff with an assessment of the different legal frameworks addressing the international waste trade. For a brief overview on the developments in international environmental law to control hazardous substances see Dupuy and Viñuales (n 2) 200 ff; see also Anne Daniel, 'Hazardous Substances, Transboundary Impacts' in Rüdiger Wolfrum (ed.), *Max Planck Encyclopedia of Public International Law* (2009) available at <www.mpepil.com> (last accessed on 30 July 2016).

remain rather vague. For example, how should materials be treated if scientific uncertainties exist with regard to the risks they imply? How should substances be regulated that involve mixtures of materials and chemical compounds? How can risks be addressed that do not derive from the physical and chemical characteristics of an end-of-life material, but stem from their unsafe and unsound handling? What about materials that are perceived as wastes by one of the trading parties, but as valuable goods by the other? Furthermore, with second-hand and used goods finding markets in less developed countries,[104] additional, ethically tinted questions are bound to arise that may challenge the acceptability of shipping wastes to another side of the world as a matter of principle.

Of course, the international trade law regime of the WTO does not provide general answers to such intricate cross-cutting issues. Recent WTO case law seems to emphasize the regulatory autonomy of its Member States. The *Brazil – Tyres* case has been pointed out as illustrating the WTO's increased acceptance of national trade measures adopted for the purpose of addressing environmental concerns. However, even though this case may have significant effects as a precedent, it is important to bear in mind that it is not binding on future adjudicating bodies. Rather, the WTO Panels and the Appellate Body have to find solutions to different challenges on a case-by-case basis – a process that will continue raising questions on the role that the WTO has in reconciling complex disputes involving both trade and 'non-trade' issues.

---

[104] See, e.g., Grosz (n 3) 86–9 with further references.

PART II

Greening the economy through waste management

6. Green Economy and sustainable development

Vera Weick[1]

EXECUTIVE SUMMARY

Significant progress has been made over the last three decades through international conferences and reports to seize the opportunities of sustainable development in view of the challenges of climate change, the limited carrying capacity of the Earth, and degrading ecosystems. In 2015, the UN General Assembly agreed on Sustainable Development Goals to guide their forward-looking Agenda 2030. Sustainable development emphasizes the enhancement of environmental, social and economic resources, with all three of them being critical to meet the needs of current and future generations.

But despite the concept's penetration into many segments of society and the rise of environmental policies throughout the world, the impact on global environmental trends has been limited. Bottlenecks in the way sustainable development has been approached in practice – with a focus on environmental protection and negative externalities – provide a basis for understanding the evolution of the Green Economy concept. In the aftermath of the last world economic crisis, the Green Economy gained attention as a concept that could overcome the connotation of environmental protection as a cost factor slowing down economic development and bring the environment and the economy into a positive relationship, in which the environment becomes an opportunity rather than a constraint, and a new driving force for economic development. Sustainability remains the vital long-term goal, but the Green Economy is describing a pathway to sustainable development. To put emphasis on the importance of including social aspects, the concept of the Green Economy has evolved and many organizations now refer to an 'inclusive Green Economy'.

---

[1] The views expressed therein are those of the author and do not necessarily reflect the views of the United Nations.

As a key feature, the Green Economy promotes investments in specific areas – also broadly referred to as green sectors – which either restore and maintain natural resources or increase efficiency in their use. These investments can lead, as any other public investment, to the creation of jobs, generation of income and development of new markets but with less emissions, resource degradation and environmental pollution. While each country has its own national conditions and the design of a Green Economy and related policies will vary, key characteristics for the process of 'greening' can be described by: (i) an increase in the share that 'green sectors' contribute to the Gross Domestic Product as well as in a country's population that is employed in these sectors; (ii) decoupling of economic growth from resource use and environmental impact; (iii) an increase in public and private investment going into green sectors; and (iv) a changing composition of aggregated consumption in which the share of environmentally friendly products and services increases.

Building on UNEP's report 'Towards a Green Economy', areas of policy-making which provide key enabling conditions for a Green Economy transition include: (i) promoting investment and spending in areas that stimulate a Green Economy (e.g. in technology, infrastructure or infant industries); (ii) limiting government spending in areas that deplete natural capital through a reduction of environmentally harmful subsidies; (iii) establishing sound regulatory frameworks that create rights, incentives, minimum standards and prohibit the most harmful forms of behaviour and substances; (iv) addressing environmental externalities and existing market failures by employing taxes and market-based instruments that promote green investment and innovation; and (v) strengthening international governance in areas where international and multilateral mechanisms regulate economic activity in addition to national laws. Depending on their current level of development, countries have different capacities to initiate and implement policy reform and cope with transformative change. Other supporting actions are therefore needed to increase capacity and strengthen institutions, provide training and skill enhancement to the workforce, and improve general education on sustainability.

6.1   FROM SUSTAINABLE DEVELOPMENT TO A GREEN ECONOMY ON THE INTERNATIONAL AGENDA

The concept of 'sustainable development' emerged on to the global stage in late 1980s out of the recognition that with a growing world population, development opportunities are threatened by the depletion of natural resources and the degradation of ecosystems. A report by the Club of Rome in 1972 analysed in different scenarios the consequences of the interactions between the Earth and human systems using five main variables: world population; industrialization; pollution; food production; and resource depletion. In some of the predictions, it saw a growing world population in a limited environment reaching the limit of its carrying capacity in the twenty-first century.[2] Mindful of these messages, the World Commission on Environment and Development (also called the 'Brundtland Commission') linked, in its 1987 report, environmental action and poverty reduction to the concept of sustainable development. It provided the most commonly used definition that describes sustainable development as 'development that meets the needs of the present without compromising the ability of future generations to meet their own needs'.[3]

This report helped to set the stage for the 1992 United Nations Conference on Environment and Development (UNCED, also known as the Earth Summit) in Rio de Janeiro, Brazil, and the Rio Declaration on Environment and Development, which established the importance of sustainable development at the international level.[4] Agenda 21, which was adopted by the Conference, called upon countries to make sustainable development a priority project of the international community, and highlighted key areas for action. Under its social and economic dimension, it highlighted the need for poverty elimination; changing consumption patterns; promotion of human health and sustainable human settlements; more sustainable population dynamics; and the integration of environment and development into decision-making. For the conservation and management of natural resources, it called for the protection of the atmosphere and fragile environments; conservation of biological diversity; an integrated approach to planning and management of resources; pollution control; and management of biotechnology and radioactive waste. Another important aspect of Agenda 21 was the strengthening of major groups, including the roles of children, women,

---

[2] Dennis Meadows et al, *The Limits to Growth – A Report to the Club of Rome* (Universe Books 1972).

[3] United Nations, 'Our Common Future', Report of the World Commission on Environment and Development, World Commission on Environment and Development. Annex to General Assembly document A/42/427 (1987) Part I.2.

[4] United Nations Conference on Environment and Development, Rio Declaration on Environment and Development, General Assembly A/CONF.151/26 (Vol. I).

non-governmental organizations, local authorities, business and industry, workers, indigenous people and farmers.[5]

In 2002, the World Summit on Sustainable Development was held in Johannesburg, South Africa, on the tenth anniversary of the Earth Summit, where governments reaffirmed their commitment to sustainable development and further elaborated the concept. The Johannesburg Declaration refers to the 'mutually reinforcing pillars of sustainable development – economic development, social development and environmental protection'.[6]

The emphasis on three pillars builds on the understanding that 'sustainability' relates to the maintenance and enhancement of environmental, social and economic resources, with all three of them being critical in order to meet the needs of present and future generations:

- *Environmental sustainability* requires that natural capital remains intact. The extraction of renewable resources should not exceed the rate at which they are renewed, and the absorptive capacity of the environment to assimilate wastes should not be exceeded. The extraction of non-renewable resources should be minimized and should not exceed agreed minimum strategic levels.
- *Social sustainability* requires that the cohesion of society and its ability to work towards common goals be maintained. Individual needs, such as those for health and well-being, nutrition, shelter, education and cultural expression should be met.
- *Economic sustainability* occurs when development, which moves towards social and environmental sustainability, is financially feasible.[7]

In 2012, 20 years after the Earth Summit, governments convened again in Rio de Janeiro, Brazil, for the United Nations Conference on Sustainable Development, also commonly referred to as the Rio+20 Conference. In its outcome document – the Rio Declaration on 'The Future We Want' – sustainable development is brought into context with

---

[5] United Nations Earth Summit, 'Agenda 21. The United Nations Programme of Action from Rio', available at <www.un.org/esa/dsd/agenda21/> (last accessed on 13 August 2015).

[6] United Nations, Johannesburg Declaration on Sustainable Development, A/CONF.199/20, Chapter 1, Resolution 1 (Johannesburg, September 2002) 1 para. 5.

[7] Richard Gilbert, Don Stevenson, Herbert Girardet and Richard Stren, *Making Cities Work: The Role of Local Authorities in the Urban Environment* (Earthscan 1996) 11–12.

Green Economy. As governments took note of the uneven progress over the last 20 years, they renewed their commitment to sustainable development and considered 'Green Economy in the context of sustainable development and poverty eradication as one of the important tools available for achieving sustainable development (. . .)'.[8]

Beyond the recognition of Green Economy, the Rio+20 Conference made an important step to bring sustainability into the centre of the goals and target set by the United Nations General Assembly by mandating the development of Sustainable Development Goals (SDGs).[9] Fifteen years after governments agreed with the Millennium Declaration to work towards the Millennium Development Goals (MDGs),[10] the 68th session of the UN General Assembly adopted in September 2015 a set of 17 Sustainable Development Goals with 169 targets.[11] SDGs cover sustainable development across its different dimensions ranging from ending hunger and poverty, health, inclusiveness, access to energy and sustainable economic growth and industrialization, to the protection of the climate, the planet's terrestrial ecosystems and oceans, emphasizing for all efforts the cross-cutting importance of peace and justice and global partnerships. While the MDGs still specified 'ensuring environment sustainability' as a specific goal, among seven others, with the SDGs sustainability has become the overall framing concept in the forward going plan – the 2030 Agenda – adopted by 193 governments.

6.2   GLOBAL TRENDS IN SUSTAINABLE DEVELOPMENT

Since its first framing in the 1970s and 1980s, the concept of sustainable development has had significant traction throughout all important segments of society, including government, business, civil society and academia, which have all responded to the challenge of sustainability to some extent. Almost every country in the UN has established a ministry or department tasked with environmental policy; and regional and local

---

8   United Nations, 'The Future We Want'. Outcome document of the World Conference on Sustainable Development (Rio+20) General Assembly A/RES/66/288 (United Nations 2012), paras 1, 12, 19, and 56.
9   Ibid., paras 245–249.
10   United Nations, 'United Nations Millennium Declaration', General Assembly A/Res/55/2 (United Nations 2000).
11   United Nations, 'Transforming our World: the 2030 Agenda for Sustainable Development', General Assembly, A/Res/70/1 (United Nations 2015) paras 54–59.

governments have also increased their capacity for implementation. The body of environmental policy has grown steadily – at the international, national and local levels – and international environmental agreements in different areas (such as biological diversity, climate change, wetland, chemicals, and hazardous waste) have driven international consensus among countries to act on specific global threats. Mainstreaming of sustainability into policies as well as the development of specific sustainability policies has become an important area of policy-making.[12]

Numerous civil society groups and research institutes have made it their main purpose to advocate and research sustainable development, and public awareness of environmental and social issues are in many cases much better developed. In the private sector, sustainability has become a central element in corporate social responsibility and many companies issue sustainability reports.[13] With the adoption of the Sustainable Development Goals, and efforts of governments to align national development strategies with the 2030 Agenda, this trend is likely to be further strengthened and accelerated.

But despite the concept's penetration into many segments of society and the rise of environmental policies throughout the world, a decade and a half into the new millennium, the impact on global trends in resource depletion, ecosystem degradation, waste generation, or greenhouse gas (GHG) emissions and the related risk of climate change, has been limited. These trends have been summarized by a variety of publications released over the last ten years, which monitor different elements of the global environment.

● The Millennium Ecosystem Assessment documented in 2005 that approximately 60 percent of the major ecosystem services it examined are being degraded or used unsustainably.[14]

---

[12]   The progress made in sustainable development at the regional, national and subnational and local levels has been recognized in the outcome document of the Rio+20 Conference, which notes that efforts to achieve sustainable development have been reflected in regional, national and subnational policies and plans, as well as the legislation, institutions, international, regional and subregional agreements and commitments. See United Nations, 'The Future We Want' (n 8) para. 22.

[13]   IUCN, 'The Future of Sustainability – Rethinking Environment and Development in the 21st Century' (World Conservation Union 2006) 2 ff, available at <http://cmsdata.iucn.org/downloads/iucn_future_of_sustainability.pdf> (last accessed on 1 August 2015).

[14]   Millennium Ecosystem Assessment, 'Ecosystems and Human Well-Being: Current State and Trends – Synthesis Report' (Island Press 2005) 1, available at <www.

- The 2011 Global Environmental Outlook *'Keeping Track of our Changing Environment'* highlighted that the global forest area has decreased by 300 million hectares since 1990 and that biodiversity in the tropics has declined by 30 percent since 1992.[15]
- The reports by the Intergovernmental Panel on Climate Change (IPCC) continue to point to the human influence on the climate system. CO_2 concentrations have increased by 40 percent since pre-industrial times, primarily from fossil fuel emissions and secondarily from net land use change emissions, leading to warming of the atmosphere and the oceans, a decrease in the amount of snow and ice, and a rise in sea levels.[16]
- The Global Footprint Network (GFP) established that humanity uses the equivalent of 1.5 of the planet's bio-capacity, meaning it takes the Earth one year and six months to regenerate what we use in a year. With current trends, it will be the equivalent of two times the Earth's bio-capacity by 2030.[17]

The reports by the International Resource Panel (IRP) recorded some progress in the relative per capita decoupling of resource use from economic development over the last 40 years, but these improvements have been overwhelmed by an absolute increase in the amounts of materials and fossil energy being used globally.[18] This trend is likely to continue, with a projected world population of 8.2 billion and an estimated 2–3 billion additional middle class consumers added to the world market by 2030.[19]

---

millenniumassessment.org/documents/document.356.aspx.pdf> (last accessed on 1 August 2015).

[15] UNEP, 'Keeping Track of our Changing Environment – From Rio to Rio+20 (1992–2012)' (UNEP 2011) 37 and 45.

[16] IPCC, 'Climate Change 2013: The Physical Science Basis'. IPCC Working Group I Contribution to the Fifth Assessment Report (AR5), Intergovernmental Panel on Climate Change (IPCC 2013) 4 ff.

[17] Global Footprint Network, 'Living Planet Report' (Global Footprint Network 2011) 9. August 19 was Earth Overshoot Day 2014, marking the date when humanity exhausted nature's budget for the year. See Global Footprint Network, 'Earth Overshoot Day', available at <www.footprintnetwork.org/en/index.php/gfn/page/earth_overshoot_day/> (last accessed on 1 August 2015).

[18] UNEP, 'Decoupling natural resource use and environmental impacts from economic growth', Summary report, International Resource Panel (UNEP 2011) 18.

[19] United Nations Department of Economic and Social Affairs, 'World Population Prospects: The 2012 Revision, Highlights and Advance Tables' (UN DESA 2012) xv, available at <http://esa.un.org/unpd/wpp/Documentation/pdf/WPP2012_HIGHLIGHTS.pdf> (last accessed on 13 August 2015); OECD, 'The

Figures on social development, while having improved in some areas, also show that in the current generation not all basic needs are being met, and – taking into account the above mentioned environmental trends – will remain a challenge for the next generation.

The 2015 Report on the Millennium Development Goals indicates significant progress against health and social indicators (such as HIV and malaria infections, global maternal mortality ratio, primary education, and deaths of children under five), and a 47 percent reduction between 1990 and 2015 of the people living on less than $1.25 a day, but, nevertheless, 836 million people still live in extreme poverty. 12.9 percent of people in developing regions remain undernourished, 663 million people across the world remain without access to improved drinking water and an estimated 2.4 billion do not have access to improved sanitation. [20]

A large portion of the improvement in global poverty reduction can be associated with fast growing countries, like China, while other countries and regions have not seen the same development. In 2012, 42 percent of the population in Sub-Saharan Africa lived on less than $1.9 a day. [21] There are 1.2 billion people who currently have no access to electricity, 95 percent of those living in Sub-Saharan Africa and developing Asia, and 2.7 billion who rely on traditional use of biomass for cooking causing harmful indoor air pollution.[22]

The 2015 MDG report states the fact that climate change and environmental degradation undermine the progress achieved. Altering ecosystems and weather patterns, together with loss of forests, overexploitation of marine fish stocks, and water scarcity, directly affect poor people whose

---

Challenges for Social Cohesion in a Shifting World', in OECD, *Perspectives on Global Development 2012: Social Cohesion in a Shifting World* (OECD 2011) 103.

[20]   United Nations, 'The Millennium Development Goals Report 2015' (United Nations 2015) 4, available at <www.un.org/millenniumgoals/2015_MDG_Report/pdf/MDG%202015%20rev%20(July%201).pdf> (last accessed on 14 March 2016); World Health Organization, 'Water Sanitation Health, Key facts JMP 2015 Report', 2, available at <www.who.int/water_sanitation_health/publications/JMP–2015-keyfacts-en-rev.pdf> (last accessed on 14 March 2016).

[21]   World Bank, 'World Bank Poverty and Equity Data', available at <http://povertydata.worldbank.org/poverty/region/SSA> (last accessed on 14 March 2016).

[22]   International Energy Agency, 'World Energy Outlook 2015' (IEA 2015) 1, available at <www.worldenergyoutlook.org/resources/energydevelopment/> (last accessed on 14 March 2016).

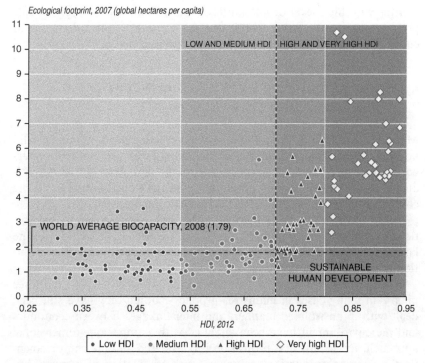

Source: UNDP, 'Human Development Report 2013, The Rise of the South: Human Progress in a Diverse World' (UNDP 2013) 35, available at <http://hdr.undp.org/sites/default/files/reports/14/hdr2013_en_complete.pdf> (last accessed on 1 August 2015).

Figure 6.1   Ecological footprint and human development of 151 countries

livelihoods are more directly tied to natural resources and who often live in vulnerable areas.[23]

A correlation between the Human Development Index and the Ecological Footprint per capita illustrates the current dilemma of countries to improve human well-being while also ensuring sustainable use of resources within the Earth's bio-capacity. As shown in Figure 6.1, only a few countries come close to creating a high level of human development without exerting unsustainable pressure on the planet's ecological resources. Sustainable development means to reach the right-hand lower

---

[23]   United Nations, 'The Millennium Development Goals Report 2015' (United Nations 2015) 8, available at <www.un.org/millenniumgoals/2015_MDG_Report/pdf/MDG%202015%20rev%20(July%201).pdf> (last accessed on 14 March 2016).

quarter with high level of human development and low ecological footprint per capita. To move to this quarter, a more significant change in current patterns of production and consumption is needed.[24]

6.3 BOTTLENECKS IN THE APPROACH TO SUSTAINABLE DEVELOPMENT

Many reasons can be put forward why sustainable development did not have the anticipated impact at the global level, ranging from international politics and the discussion around 'common but differentiated responsibilities' among developing and developed countries to the commonly cited lack of funds, technology and capacity.[25]

Only a few points are highlighted in what follows, describing some bottlenecks in the way sustainable development has been approached over the last two decades and which provide a basis for understanding the evolution of the Green Economy concept.

While sustainable development has been established around the three pillars, in many cases in mainstream policy-making they have still been dealt with in an isolated manner, and inter-linkages between economics and the environment have been analysed with a rather one-dimensional approach. Environmental protection is considered an important element on the road to economic development and eradication of poverty, but it has also been perceived as a burden, which may slow down economic growth. Economic growth is seen as the main driver to lift people out of poverty and create jobs and income for the growing world population. Some extent of environmental degradation has been perceived as inevitable in the course of development.[26]

---

[24] UNDP, 'Human Development Report 2013, The Rise of the South: Human Progress in a Diverse World' (UNDP 2013) 35, available at <http://hdr.undp.org/sites/default/files/reports/14/hdr2013_en_complete.pdf> (last accessed on 1 August 2015).

[25] Beijing Normal University and UNEP, 'Green Economy: Theory, Methods and Cases from the United Nations' Perspective' (Chinese version) (UNEP 2015). The concept of common but differentiated responsibilities was established in Principle 7 of the 1992 Rio Declaration. It recognizes historical differences in the contributions of developed and developing countries to global environmental problems, and differences in their respective economic and technical capacity to tackle these problems. See also UN, Rio Declaration (n 4).

[26] Ibid. Building on the hypothesis established in the Environmental Kuznet Curve, describing the relationship between environmental quality and economic development over time in a U-shape, environmental degradation would get worse

On the other hand, economic activity has mainly been described in terms of the negative impacts on the environment – e.g. through the concept of negative externalities[27] – which need to be addressed through policy instruments. Policies adopted to solve environmental problems have then tended to be reactive and linked to reduction targets. While this approach has been effective in addressing specific problems, as for example in the phasing out of harmful substances, it has not been able to affect the way economic and social policies are developed in a more fundamental manner.[28]

6.4   FROM CRISIS TO OPPORTUNITY – THE EMERGENCE OF THE GREEN ECONOMY CONCEPT

Against this background – and in the aftermath of the recent world economic crisis – Green Economy gained attention as a concept that could overcome the negative connotation of environmental protection as a cost factor slowing down economic development, and bring the environment and the economy into a positive relationship, in which the environment becomes an opportunity rather than a constraint and a new driving force for economic development.

In the period 2008–10, the world experienced the worst global economic recession since the Great Depression of the 1930s. In the decade before the crisis, it had seen unprecedented economic growth driven by the accumulation of financial capital and world trade. The systemic risks of this development became evident when the crisis struck the financial services sector. The subsequent global economic slowdown wiped out a large portion of global wealth, spurred unemployment and consigned millions of people in developing economies back to poverty. In the same period, the

---

over the course of development until average income reaches a certain point, after which environmental quality improves with increasing incomes. This relationship could empirically only be proved for some pollutants (e.g. sulphur dioxide and lead) but not for many others indicators (e.g. not for GHG emissions, land and resource use); David I. Stern, 'The Rise and Fall of the Environmental Kuznets Curve' (2004) 32(8) *World Development* 1419. But the basic idea of 'grow first, clean up later' prevailed in development policies.

[27]   Ibid. In economics, an externality is a cost or benefit that results from an activity and that affects an otherwise uninvolved party who did not choose to incur that cost or benefit, see James Buchanan and William Craig Stubblebine, 'Externality' (1962) 29(116) *Economica* 371.

[28]   Beijing Normal University and UNEP, ibid.

world experienced a peak in fuel prices, and a related increase in food and commodity prices, putting an additional burden on poorer segments of society and leading to social unrest in over 20 countries around the world.

The manifestation of the concurrent crises led to a reconsideration of the traditional development patterns observed in the last decades, which have prioritized investments in physical capital (e.g. infrastructure), human capital (e.g. employment) and financial capital, with the aim of increasing economic growth. In contrast, relatively small amounts were invested in environmental capital or resource efficiency, e.g. in areas such as renewable energy, energy efficiency, public transportation, sustainable agriculture, ecosystem and biodiversity protection, or land and water conservation.[29]

As a result of the global market and financial crises, these investments became the centre of attention in a report released in early 2009 calling for a 'Global Green New Deal (GGND)' to restore the economy, reduce poverty, and reduce carbon emissions and the degradation of ecosystems. Inspired by the 'New Deal' – a government-led investment plan during the Great Depression – the document proposed a framework for green stimulus programmes as well as supportive domestic and international policies, including support to Least Developed Countries. In a nutshell, the GGND called for 1 percent of global GDP to be invested within two years in green infrastructure to support growth of the economy, including in energy-efficient buildings, sustainable transport, renewable energy, sustainable agriculture and water management.[30]

The underlying notion of 'investments in the environment', presented in the GGND as a driving force for economic recovery, is a central feature of the concept of Green Economy. Investments in specific areas – also broadly referred to as green sectors – which either restore and maintain

---

[29]   UNEP, 'Towards a Green Economy: Pathways to Sustainable Development and Poverty Eradication', Synthesis Report (UNEP 2011) 1–2, available at <www.unep.org/greeneconomy/Portals/88/documents/ger/GER_synthesis_en.pdf> (last accessed on 1 August 2015).

[30]   Edward Barbier, *A Global Green New Deal: Rethinking the Economic Recovery* (Cambridge University Press and UNEP 2010). UNEP analysed, in a 2009 update on the GGND for the Pittsburg G20 Summit, the stimulus packages for seven countries (China, France, Germany, the United States, Mexico, Republic of Korea and South Africa). China and South Korea stood out with 34 and 78 percent, respectively, of green stimulus, other countries having green components in their stimulus packages ranging between 10 and 20 percent, see UNEP, 'Global Green New Deal. An update for the G20 Pittsburg Summit' (UNEP 2009) 2, available at <www.unep.ch/etb/publications/Green%20Economy/G%2020%20policy%20brief%20FINAL.pdf> (last accessed on 1 August 2015).

natural resources or increase efficiency in their use can lead, as with any other public investments, to the creation of jobs, generation of income and development of new markets. This, however, is achieved with less emissions, resource degradation and environmental pollution, and, as such, provides an alternative to the perceived inevitable trade-off between economic development and environmental quality.[31]

The term 'Green Economy' first appeared in 1989 in the report Blueprint for a Green Economy which did not provide a clear definition but elaborated a few basic concepts. It noted that the interdependence between the environment and the economy is key to understanding the concept of sustainable development. 'The environment must be seen as a valuable, frequently essential input to human wellbeing', and '[s]ustainable development means a change in consumption patterns towards environmentally more benign products, and a change in investment patterns towards augmenting environmental capital'.[32]

6.5   DEFINING GREEN ECONOMY AND ITS RELATION TO SUSTAINABLE DEVELOPMENT

A definition, which was commonly referred to in recent years, was suggested in UNEP's 2011 report 'Towards a Green Economy' which described a Green Economy as '[a]n economy that results in improved human well-being and social equity, while significantly reducing environmental risks and ecological scarcities'.[33]

This definition, anchored at the visionary level, provides guidance on the elements that need to come together in a Green Economy, building conceptually on the three pillars of sustainable development. Improving human well-being is linked to environmental improvements – less risk and scarcity – but also to social equity. Any effort to address an environmental problem comes with social considerations. Simply put, solar panels that are produced by workers in poor working conditions not receiving adequate wages cannot be considered as the right pathway. Nor could producing solar panels without a concept for recycling or safe disposal of their components after use, as this practice creates additional environmental risks although GHG emissions may be reduced.

---

[31]   UNEP, 'Towards a Green Economy' (n 29) 2–3.
[32]   David Pearce, Anil Markandya and Edward Barbier, *Blueprint for a Green Economy* (Earthscan Publications 1989) xiv.
[33]   UNEP, 'Towards a Green Economy' (n 29) 2.

A report issued by the UN Environmental Management Group (EMG) in December 2011 'Working towards a Balanced and Inclusive Green Economy' specifically highlighted the social element. The elimination of poverty and the achievement of social justice are described as direct objectives and targets of investment in an inclusive Green Economy – going beyond the idea that a Green Economy will draw mainly on the inter-linkages between environment and economy with poverty and equity issues being addressed indirectly.[34] Since then the concept of 'inclusive Green Economy' (IGE) with an emphasis on the inclusion of all segments of society and reduction of inequality of global wealth concentration, has become the reference for many UN agencies as well as other organizations.[35]

In 2015, evolving from its earlier work on Green Economy, UNEP published a summary for leaders 'Uncovering pathways towards an IGE', describing an IGE as an economy that is 'low carbon, efficient and clean in production but also inclusive in consumption and outcomes, based on sharing, circularity, collaboration, solidarity, resilience, opportunity, and interdependence'. Noting the opportunity arising with the globally adopted SDGs and 2030 Agenda 'to reframe economic policy around the core elements of sustainability', it describes core elements of the inclusive Green Economy as central to their achievement.[36]

In a more operational definition, a Green Economy is seen as one whose growth in income and employment is driven by investments that:

● Reduce carbon emissions and pollution;
● Enhance energy and resource efficiency; and
● Prevent the loss of biodiversity and ecosystem services.

---

[34]   UN Environment Management Group, 'Working Towards a Balanced and Inclusive Green Economy – A United Nations System-Wide Perspective' (UNEMG 2011) 13, available at <www.fao.org/fileadmin/user_upload/sustainability/pdf/GreenEconomy-Full.pdf> (last accessed on 1 August 2015).

[35]   Examples include the work of the Poverty Environment Network that released in 2012 a joint agency paper on 'Building an Inclusive Green Economy for All', available at <www.wri.org/sites/default/files/pdf/building_inclusive_green_economy_for_all.pdf>; or the work of UNDP, 'Examples of Inclusive Green Economy Approaches in UNDP's Support to Countries' (UNDP 2012) available at <www.undp.org/content/dam/undp/library/Environment%20and%20Energy/Examples-of-Inclusive-Green-Economy-Approaches-in-UNDP's-Support-to-Countries-June2012_Updated-Sept2012.pdf> (both last accessed on 17 March 2016).

[36]   UNEP, 'Uncovering Pathways towards an IGE' (UNEP 2015) 6 and 11, available at <www.unep.org/greeneconomy/Portals/88/documents/GEI%20Highlights/IGE_NARRATIVE_SUMMARY.pdf> (last accessed on 17 March 2016).

These include investments in human and social capital, and recognize the central position of human well-being and social equity as core goals promoted by growth in income and employment.[37]

Using this operational definition, a Green Economy can be placed alongside other major economic patterns in human history characterized by their main driver for growth and income, such as an agricultural economy, an industrial economy, a service economy, or a knowledge-based economy where the production, distribution and use of knowledge drives cross-industry growth, wealth creation and employment. Accordingly, an economy that is based on green products and services and – to use a more decisive indicator – in which more than half of the goods and services have authoritative environmental certification, could be called a Green Economy.[38]

Other entities such as the World Resources Institute (WRI) point out that Green Economy promotes a triple bottom line: sustaining and advancing economic, environmental and social well-being.[39] Or the World Conservation Union (IUCN) which emphasizes that Green Economy is complementing sustainable development with its three indispensable pillars by specifically putting focus on the sustainability of the economic pillar.[40]

The UN Department of Economic and Social Affairs (UN DESA) conducted a review of the existing literature, and concluded that various definitions of a Green Economy exist and that they are generally consistent, having sustainable development as their ultimate objective and being a means to reconcile economic development and environmental sustainability, without ignoring social aspects.[41] Sustainability remains the vital long-term goal, and the Green Economy concept does not replace the concept of sustainable development but describes a pathway which builds on the recognition that 'achieving sustainability rests almost entirely on getting the economy right'.[42]

As mentioned above in section 6.1, this is also the relationship that governments described in the outcome document of the Rio+20 Conference, in which Green Economy is recognized as an important tool to achieve sustainable development. At the same time, the document notes in the

---

[37] UNEP, 'Towards a Green Economy' (n 29) 2.
[38] Beijing Normal University and UNEP (n 25).
[39] World Resource Institute, 'Q&A: What Is a Green Economy?' available at <www.wri.org/blog/2011/04/qa-what-green-economy–0> (last accessed on 1 August 2015).
[40] International Union for the Conservation of Nature, 'What is the difference between the concept of a Green Economy and Sustainable Development?' available at <www.iucn.org/news_homepage/events/iucn___rio___20/iucn_position/green_economy/> (last accessed on 1 August 2015).
[41] UN DESA, 'World Population Prospects' (n 19).
[42] UNEP, 'Towards a Green Economy' (n 29) 2.

context of implementation of Green Economy policies by countries, 'that each country can choose an appropriate approach in accordance with national sustainable development plans, strategies and priorities'.[43] This understanding has been reaffirmed by a UNEP Governing Council Decision in 2013, which recognizes that 'there are different approaches, visions, models and tools available to each country, in accordance with its national circumstances and priorities, to achieve sustainable development'.[44]

Similarly for the SDGs, the GA resolution refers to 'universal goals and targets which involve the entire world, developed and developing countries alike', but which are 'acceptable and applicable to all, taking into account different national realities, capacities and levels of development and respecting national policies and priorities'.[45]

It is indeed very important to note that each country has its own national conditions, and Sustainable Development Goals are not the same across countries. Accordingly, the design of a Green Economy and related policies will vary.[46] However, for a better understanding of the concept, some key characteristics for the process of 'greening' as well as key enabling conditions for a Green Economy can be described.

6.6   KEY CHARACTERISTICS OF A GREEN ECONOMY

The economic output of a country or region is commonly measured by the Gross Domestic Product (GDP), an aggregated measure of production, which sums up the gross value added of all resident institutional units engaged in production – or put less technically – a measure for the market value of all final goods and services produced within a country within a time period. GDP can also be used to calculate the growth of an economy from year to year, or for shorter time periods, the pattern of which is then used to indicate economic progress, or if an economy is in recession.[47]

While GDP is often criticized for not capturing environmental

---

[43]   United Nations, 'The Future We Want' (n 8) 11, para. 59.

[44]   United Nations Environment Programme Governing Council, Decisions adopted by the Governing Council/Global Ministerial Environment Forum at its first universal session, Decision 27/8, UNEP/GC.27/17 (2013).

[45]   United Nations, 'Transforming our World: The 2030 Agenda for Sustainable Development', General Assembly, A/Res/70/1 (United Nations 2015) para. 5.

[46]   Beijing Normal University and UNEP (n 25).

[47]   United Nations, System of National Accounts (United Nations 1993) 54, paras 2.172 and 2.173.

degradation or different forms of informal labour, it still is useful as a basic measure for economic activity of a country. Problems mainly arise when GDP is used beyond economic activity to describe the human well-being of a country or overall societal progress, which it does not measure. Accordingly, the Rio+20 outcome document recognized '[t]he need for measures of progress to complement gross domestic product in order to better inform policy decisions' and different organizations and countries work on alternative indicators (see below and section 7.7).[48]

A national economy can be further disaggregated into economic sectors, by describing a country's population based upon the economic activity that it is engaged in, such as agriculture, mining, manufacturing, construction, services, or education. If described by expenditure components, GDP is described as the sum of consumption, investment, government spending, and net exports.

Taking into account these elements, a Green Economy can be characterized by an increase in the proportion of economic activity, which produces environmental goods and services and which contributes to reduced carbon emissions and pollution; increased energy and resource efficiency and conservation and better management of natural resources – activities which can be loosely referred to as activities in 'green sectors'. In a nutshell, in a Green Economy there would be an increase in the share that these 'green sectors' (e.g. public transport, sustainable construction, watershed management, etc.) contribute to the GDP as well as an increase in a country's population that is employed in these sectors.[49]

Another important element to characterize a Green Economy is the decoupling of economic growth from resource use and environmental impact, meaning a decrease in the resources and energy used per unit of economic output, as well as a reduction of environmental impact –

---

[48]   United Nations, 'The Future We Want' (n 8) para. 38.

[49]   IILS states that a green sector is specified both in relative and absolute terms, e.g. an enterprise or industry in the green sector must be relatively low-carbon-intensive compared to other industries or enterprises in the economy, and the total CO_2 emissions of the green sector as a whole must be low enough to be sustainable. IILS also notes that it is advisable not to link the green sector to specific industries, sectors, products or services. A green sector may comprise different industries in different countries. Also the exact 'face' of the green sector is not fixed and changes over time. A green industry today might not be a green industry in ten years. See International Institute for Labour Studies, 'Defining "Green" – Issues and Considerations', EC-IILS Joint Discussion Paper Series No. 10 (International Institute for Labour Studies/International Labor Organization 2011) 21, available at <www.ilo.org/wcmsp5/groups/public/---dgreports/---inst/documents/publica tion/wcms_194180.pdf> (last accessed on 1 August 2015).

emissions and pollution – per unit of economic output. This decoupling would allow a national economy to produce the same amount of goods and services, or even experience further economic growth, while using less resources and generating less emissions and pollution.[50]

Building on the expenditure elements of GDP referred to above, a Green Economy can be characterized by an increase in public and private investment going into green sectors and by a changing composition of aggregated consumption in which the share of environmental friendly products and services increases.

For an analysis of aggregate consumption in a Green Economy, it is important to take a differentiated view at countries and different levels of income, e.g. a high-income country with saturated markets or a low-income country where growth in products and services and their consumption is needed to meet people's basic needs and future aspirations. Given these different country situations, a Green Economy is not necessarily characterized by an absolute decrease in consumption, but rather focuses on its composition, seeking to decrease consumption in areas where it comes with negative externalities to the environment.

While the above highlights a few broad elements, there are efforts by different organizations to provide a more comprehensive set of indicators that can be used to measure progress towards an inclusive Green Economy. Other efforts are focused on the development of a composite indicator that informs on national or city level green economies, in terms of performance and perceptions. Due to the complexity of socio-economic and environmental systems, assessing progress with a single metric is challenging and bears the risk of misleading policy messages, if they are not well constructed or are wrongly interpreted. But frameworks of indicators are available that can be applied to countries in different regions of the world and at different stages of development, and that can be customized by government to meet their respective needs.[51]

---

[50] UNEP, 'Decoupling Natural Resource Use and Environmental Impacts from Economic Growth', Summary report, International Resource Panel (UNEP 2011) 8, available at <www.unep.org/resourcepanel/decoupling/files/pdf/decoupling_report_english.pdf> (last accessed on 1 August 2015).

[51] UNEP, 'Towards a Green Economy' (n 29) 11, 18 and 24. For more information on these indicator frameworks to measure progress towards a Green Economy, see UNEP, 'A Guidance Manual on Indicator Publication for Green Economy Indicators' (UNEP 2014) available at <www.unep.org/greeneconomy/Portals/88/documents/GEI%20Highlights/UNEP%20INDICATORS%20GE_for%20web.pdf> (last accessed on 13 August 2015); OECD, 'Towards Green Growth: Monitoring Progress' (OECD 2011) available at <www.oecd.org/greengrowth/48224574.pdf> (last accessed on 13 August 2015), and Green Growth Knowledge Platform, 'Moving

6.7   CONCEPTS RELATED TO A GREEN ECONOMY

A number of concepts are closely related to a Green Economy and can be embedded within its conceptual framework. They often have their origin within a specific area, such as resource flows, industry, and production and consumption, but have in their application by institutions evolved into more comprehensive concepts encompassing elements that are broader than the original term suggests and include objectives and characteristics similar to the Green Economy concept. Those include, among others:

Green Growth

Several institutions, including the World Bank, the Organisation for Economic Co-operation and Development (OECD), the Global Green Growth Institute (GGGI) and the United Nations Economic Commission for Asia and Pacific (UN-ESCAP), consider green economic issues under the concept of 'green growth', and several definitions have been developed for this term. According to UN-ESCAP, green growth refers to 'economic progress that fosters environmentally sustainable, low-carbon and socially inclusive development'. While growth traditionally suggests an emphasis on quantitative expansion of an economy, in this context 'growth' is not limited to output growth, but rather it is elevated to cover 'economic progress'.[52] Similarly, according to the OECD, 'green growth means fostering economic growth and development, while ensuring that natural assets continue to provide the resources and environmental services on which our well-being relies'.[53]

Circular Economy

Applying life cycle principles at national level, the concept of the circular economy, which is written into legislation in China, refers to an economy that reduces the consumption of resources and the generation of wastes,

---

Towards a Common Approach on Green Growth Indicators', GGKP Scoping Paper (Paris 2013) available at <www.oecd.org/greengrowth/GGKP%20Moving%20 towards%20a%20Common%20Approach%20on%20Green%20Growth%20 Indicators%5B1%5D.pdf> (last accessed on 13 August 2015).

[52]   United Nations Economic and Social Commission for Asia and the Pacific, Asian Development Bank and UNEP, 'Green Growth, Resources and Resilience – Environmental Sustainability in Asia and the Pacific' (UN ESCAP 2012) xv.

[53]   OECD, 'Towards Green Growth – A Summary for Policy Makers' (OECD 2011) 4.

and reuses and recycles wastes throughout the production, distribution and consumption processes. Investment in resource-efficient technologies and preventative waste management are expected to generate new sources of income and jobs, while building a resource-efficient society.[54]

Green Industry

The UN Industrial Development Organization (UNIDO) uses green industry as a term to describe 'economies striving for a more sustainable pathway of growth, by undertaking green public investments and implementing public policy initiatives that encourage environmentally responsible private investments'. Green industry promotes sustainable patterns of production and consumption, i.e. patterns which produce products that are responsibly managed throughout their lifecycle.[55]

Green Jobs

According to the International Labor Organization (ILO) Institute for Labour Studies (IILS), '[g]reen jobs are those jobs maintained or created in the transition process towards a Green Economy that are either provided by low-carbon intensive industries (enterprises) or by industries (enterprises) whose primary output function is to greening the economy'.[56] However, jobs in low-carbon or green industries are not necessarily safe and healthy jobs with adequate remuneration and social coverage. A report on 'Green jobs' jointly published by UNEP, ILO, the International Trade Union Confederation (ITUC) and the International Employers Organization (IOE) in 2008 highlights that in addition to environmental considerations, green jobs also need to reflect 'decent work'.[57]

---

[54] United Nations Environment Management Group, 'Working Towards a Balanced and Inclusive Green Economy – A United Nations System-Wide Perspective' (UN EMG 2011) 29 ff.

[55] United Nations Industrial Development Organization, 'Green Industry: Policies for Supporting Green Industry' (UNIDO 2011) 9 ff, available at <www.unido.org/fileadmin/user_media/Services/Green_Industry/web_policies_green_industry.pdf> (last accessed on 1 August 2015).

[56] International Institute for Labour Studies, 'Defining "Green"' (n 49) 22.

[57] UNEP, 'Green Jobs: Towards Decent Work in a Sustainable, Low-Carbon World' (UNEP 2008) 32 ff, available at <www.unep.org/PDF/UNEPGreenjobs_report08.pdf> (last accessed on 1 August 2015). Decent work is thereby understood as: (i) productive and secure work; (ii) that ensures respect for labour rights; (iii) provides an adequate income; (iv) offers social protection, and includes social dialogue, unions, freedom, collective bargaining and participation.

Sustainable Production and Consumption

The Johannesburg Plan of Implementation agreed to by governments at the World Summit on Sustainable Development in 2002 specifically highlighted the concept of sustainable consumption and production (SCP).[58] In a most widely used definition, 'SCP is a holistic approach to minimizing the negative environmental impacts from consumption and production systems while promoting quality of life for all.'[59] SCP encompasses a wide range of tools and approaches ranging from waste management, through cleaner production and sustainable transport, to eco-labelling and certification as well as sustainable public procurement and sustainable lifestyles.

Green Accounting

Green accounting aims to address the weaknesses of conventional economic accounting and related indicators such as gross domestic product, of not capturing the priceless environmental and social externalities. It aims to incorporate the amount of natural resources used and pollutants expelled into economic accounting in order to provide a detailed measure of all environmental consequences of economic activities. The System of Environmental-Economic Accounting (SEEA) by the UN Statistical Commission provides the internationally agreed standards, concepts, definitions, classifications, accounting rules, and tables for producing internationally comparable statistics on the environment and its relationship with the economy.[60]

Beyond the approaches that international organizations are applying in their work, certain governments have identified concepts which are also related to a Green Economy and include similar features, such as a

---

[58] United Nations, Plan of Implementation of the World Summit on Sustainable Development, A/CONF.199/20, Johannesburg (2002) Chapter 1, Resolution 2 paras 13 and 14.

[59] UNEP, 'Paving the Way for Sustainable Consumption and Production: the Marrakech Process Progress Report' (UNEP 2011) 2, available at <www.unep. org/10yfp/Portals/50150/downloads/publications/Paving_the_way/Paving_the_way_final.pdf> (last accessed on 1 August 2015).

[60] United Nations, 'System of Environmental and Economic Accounting 2012. Central Framework' (United Nations 2014) vii, available at <http://unstats. un.org/unsd/envaccounting/seeaRev/SEEA_CF_Final_en.pdf> (last accessed on 1 August 2015).

'sufficient economy' in Thailand, 'ecological civilization' used in China, or 'Vivir Bien (Living well)' used in Bolivia.[61]

6.8   KEY ENABLING CONDITIONS FOR A GREEN ECONOMY

To facilitate the transition to a Green Economy, governments play a key role in providing the enabling conditions for a shift of investment using targeted public expenditures, policy reforms and changes in regulation. UNEP's 2011 Report 'Towards a Green Economy' notes that 'with the right mix of fiscal measures, laws, norms, international frameworks, know-how and infrastructure in place, then the Green Economy should emerge as a result of general economic activity'. It further describes enabling conditions 'as conditions that make green sectors attractive opportunities for investors and businesses'.[62] A large part of the measures that support a transition to a Green Economy, is indeed focused on creating and maintaining conditions so that private actors will have incentives to invest in green economic activity; however, a leading role remains with governments to develop the broader policy frameworks. 'Towards a Green Economy' highlights five key areas of policy-making as creating the enabling conditions that support a Green Economy transition.[63]

Promoting Investment and Spending in Areas that Stimulate a Green Economy:

This includes: (i) the promotion of innovation in new technologies and behaviours; (ii) investments in common infrastructure; and (iii) public support for infant green industries. Typical examples of this are: (a) subsidies to basic research in universities or applied research in labs and industry; (b) investment in low-carbon public transport; or (c) subsidies – price-support measures, tax incentives, direct grants or loan support – for generation of renewable energy, e.g. through feed-in tariffs.[64]

---

[61]   UNEP, 'Multiple Pathways to Sustainable Development – Initial Findings from the Global South' (UNEP 2015) describes these different national pathways towards an inclusive Green Economy.

[62]   UNEP, 'Towards a Green Economy' (n 29) 552.

[63]   Ibid., 552 and 553.

[64]   Ibid., 555. A feed-in tariff is a policy instrument that makes it mandatory for energy companies or utilities responsible for operating the national grid to purchase electricity from renewable energy sources at a predetermined price that is

Limiting Government Spending in Areas that Deplete Natural Capital:

This can be achieved through a reduction of environmentally harmful subsidies. While, as noted above, there are legitimate reasons for using subsidies, in other cases they can be harmful to the environment and present a significant economic and environmental cost to countries. At the same time, they reduce the profitability of green investments by giving wrong price signals. Typical examples for this are subsidies provided in the fisheries sector, which lead to overfishing, or fossil fuel consumption subsidies, which increase their use and reduce the incentive for firms and consumers to adopt energy efficiency measures.[65] Global subsidies for fossil fuels and nuclear power are estimated to range between USD 544 billion and USD 1.9 billion, depending on how a subsidy is defined. This remains significantly higher than financial support to renewables (see section 6.9).[66]

Establishing Sound Regulatory Frameworks

Instituting sound regulatory frameworks can: (i) create rights and incentives; (ii) remove barriers to green investment; (iii) create minimum standards; or (iv) prohibit certain activities entirely to regulate the most harmful forms of behaviour and substances. Typical examples are energy efficiency standards for products, or property laws and access rights related to water, agriculture, forests and fisheries, which encourage the sustainable use of a resource. But regulation to make the provision of certain information mandatory – e.g. through labels – can influence the decisions of consumers and investors.[67]

Addressing Environmental Externalities and Existing Market Failures:

This can be achieved by employing taxes and market-based instruments that promote green investment and innovation. Via a corrective tax,

---

sufficiently attractive to stimulate new investment in renewable energy, see UNEP, 'Feed-in tariffs and a policy instrument for promoting renewable energies and green economies in developing countries' (UNEP 2012) available at <www.unep. org/pdf/UNEP_FIT_Report_2012F.pdf> (last accessed on 1 August 2015).

[65] UNEP, 'Towards a Green Economy' (n 29) 561 and 562.

[66] Ren21, 'Global Status Report, Key findings 2014', Renewable Energy Policy Network for the 21st Century (2014) 12, available at <www.ren21.net/Portals/0/documents/Resources/GSR/2014/GSR2014_KeyFindings_low%20res.pdf> (last accessed on 1 August 2015).

[67] UNEP, 'Towards a Green Economy' (n 29) 564 and 565.

charge or levy, or other market-based instruments such as tradable permit schemes, a negative externality – such as pollution or health impacts – can be incorporated in the price of a good or service. The increase in price then provides an incentive to reduce emissions, use a resource more efficiently and stimulate innovation. Typical examples are road-charging schemes, levies on natural resource extraction, license-based fees for fisheries, or cap-and-trade schemes such as the Kyoto Protocol for GHG emissions. In the case of ecosystems, where markets are often completely lacking, 'payment for ecosystem services' schemes help to create markets by asking for compensation for providing services such as carbon sequestration, watershed protection, or landscape beauty.[68]

Strengthen International Governance:

In areas where international and multilateral mechanisms regulate economic activity in addition to national laws, Multilateral Environmental Agreements (MEAs) such as the Montreal Protocol on Substances that Deplete the Ozone Layer, the Basel Convention on the Control of Transboundary Movements of Hazardous Wastes and their Disposal, or the United Nations Framework Convention on Climate Change (UNFCCC) play a significant role in promoting green economic activity by establishing legal and institutional frameworks for addressing global environmental challenges. The UNFCCC's Kyoto Protocol has already stimulated growth in a number of economic sectors, and UNFCCC's recent Paris Agreement encouraging countries set National Determined Contributions (NDCs) holds much potential to influence a transition to a Green Economy. But also the international trading scheme provide clues to accelerating the transition, as in the negotiations around the removal of fisheries subsidies or liberalization of the agricultural markets.[69]

In line with the emphasis put on countries' taking different approaches in the Rio+20 outcome document, it is important to note again that the Green Economy strategies, the mix of policy tools and the time frames for their implementation will vary from country to country, and it is not possible or advisable to develop a single policy mix that is applicable to all countries. Depending on their current level of development, countries may also have different capacities to initiate and implement policy reform and

---

[68]   Ibid., 557 ff. Payments for ecosystem services are incentives offered to farmers or landowners in exchange for managing their land to provide some sort of ecological service. Millennium Ecosystem Assessment, 'Ecosystems and Human Well-Being' (n 14) provided a comprehensive overview of these services.

[69]   Ibid., 563 ff.

cope with transformative change. Other supporting actions are therefore needed to increase capacity and strengthen institutions, provide training and skill enhancement to the workforce and improve general education on sustainability.[70]

A recent study by the Green Growth Best Practice (GGBP) initiative compiling lessons learned from country experiences highlights two important economy-wide policies as foundations for green growth: (i) green innovation policy which supports the development of 'breakthrough' technologies and business models; and (ii) labour market and skills development policies to overcome bottlenecks to investment, increase employment opportunities, smooth the transition of workers from declining sectors, and reduce social inequality especially for marginalized and lower skill workers. [71]

6.9　ENABLING CONDITIONS IN ACTION – RENEWABLE ENERGIES

One sector that has seen major developments worldwide with respect to the creation of enabling conditions is the renewable energy sector, and the trends and related impacts are well monitored. There has been a steady increase in use of renewable energy over the last decade. As examples, solar photovoltaic (PV) total capacity increased worldwide from 3.7 gigawatts in 2004 to 139 gigawatts in 2013, and total wind power capacity from 17 gigawatts in 2000 to 318 gigawatts in 2013. Renewable energy provided an estimated 19 percent of global final energy consumption in 2012 and continued to grow in 2013.[72]

This development is supported by an evolving policy landscape, in countries across the globe, which confirms suitability and adoptability of measures for different regions and countries – from low-income to high-income countries. The 2014 REN21 'Global Status Report' states that:

> [b]y early 2014 at least 144 countries had renewable energy targets and 138 countries had renewable energy support policies in place (. . .). Developing and

---

[70]　Ibid., 553, 570 and 571.

[71]　Green Growth Best Practice Initiative, 'Green Growth in Practice, Lessons Learned from Country Experiences – Executive Summary' (2014) 13, available at <www.ggbp.org/sites/all/themes/ggbp/uploads/Green-Growth-in-Practice-062014-ES.pdf> (last accessed on 1 August 2015).

[72]　REN21 (n 66) 5, 17 and 19. The 19 percent from renewables includes 9 percent traditional biomass, 4.2 percent biomass/geothermal/solar heat, 3.8 percent hydropower, 1.2 percent wind/solar/biomass/geothermal power, and 0.8 percent biofuels.

emerging economies have been leading the expansion in recent years, up from 15 countries that had introduced measures in 2005 to 95 in 2013.[73]

Global new investment in renewable power and fuels peaked in 2011 at USD 279 billion, and then declined slightly in 2012 and 2013, which can partly be explained by the uncertainty over incentives policies in Europe and the US but also by a sharp reduction in technology costs. Despite a decline in investment in solar PV of 22 percent in 2013, levels of new installations are still increasing (up from 100 to 139 gigawatt in 2013). At the same time, renewable energy was estimated to provide 6.5 million jobs globally in 2013.[74]

6.10   A GLOBAL GREEN ECONOMY INVESTMENT SCENARIO

This chapter concludes by summarizing an important piece of analysis from UNEP's 2011 Report 'Towards a Green Economy' which aimed to test the underlying hypothesis of a Green Economy: that investing in the environment delivers positive macroeconomic results, in addition to improving the environment. The report provided a comprehensive overview of the opportunities and challenges and enabling conditions in sectors that are key for a transition to a Green Economy, including agriculture, fisheries, water, forests, renewable energy, manufacturing, waste, building, transport, and tourism. It further analysed the impact that a shift of investment in these sectors could have over a 40-year time horizon at the global level.

For this, the report used a system dynamic model – the 'Threshold 21 (T21) World Model', which is one of the most advanced in terms of considering the economic, social and environmental variables that influence sustainable development in an integrated manner. The T21 model applied for this global modelling exercise comprised sectoral models for the above-mentioned sectors integrated into a global model.

In a nutshell, the analysis projects the short-term, medium-term and long-term impact of investing 2 percent of global GDP on a yearly basis between 2010 and 2050 in specific green economic activities across these key sectors, e.g. by increasing renewable energy generation, by expanding conservation agriculture or by curbing forestation. At the 2010 level, the 2 percent of global GDP amounted to about USD 1.3 trillion per year.

---

[73]   Ibid., 6. By type, renewable energy policies include feed-in tariffs, tendering, Renewable Portfolio Standards (RPS)/quota, net metering, heat obligation, biofuel blend mandate, with feed-in tariffs being the main driver for change.

[74]   Ibid., 9, 17 and 20 ff.

Table 6.1 Comparison of scenarios for selected sectors and objectives

Sector and objective	BAU Scenarios[a]	Green Scenarios
Agriculture Yield increase	Higher utilisation of chemical fertilisers	Expansion of conservation agriculture, using organic fertilisers, among others
Energy Expansion of power generating capacity	Thermal generation (fossil fuels)	Renewable energy power generation
Fisheries Increase production	Expansion of the vessel fleet, pushing catch in the short-term	Reduction of the vessel fleet, investing in stock management to increase catch in the medium- and longer-term
Forestry Increase production	Increase deforestation	Curb deforestation and invest in reforestation (expanding planted forests)
Water Manage supply and demand	Increase water supply through higher withdrawal	Invest in water efficiency measures, water management (including ecosystem services) and desalination

Note: [a] Refers to BAU1 and BAU2 with additional investments allocated to match existing patterns.

Source: UNEP, 'Towards a Green Economy: Pathways to Sustainable Development and Poverty Eradication', Synthesis Report (UNEP 2011) 1–2, available at <www.unep. org/greeneconomy/Portals/88/documents/ger/GER_synthesis_en.pdf> (last accessed on 1 August 2015).

About 50 percent of this is allocated to the development of renewable energy sources and energy efficiency, particularly in buildings, industry and transport. The remainder is going to improve waste management, public transport infrastructure, and a range of natural capital-based sectors, including agriculture, fisheries, forestry, tourism and water supply.

This is referred to as a 'Green Economy scenario'. The Green Economy scenario is then compared with a 'business as usual scenario' (BAU), in which the same amount is invested over the same period simply replicating historical trends and assuming no fundamental changes in policy or external conditions to alter the trends (see examples in Table 6.1).[75]

---

[75] This is based on the assumption that current trends will continue, with only minor progress shifting to a Green Economy (e.g. high-energy use and emissions

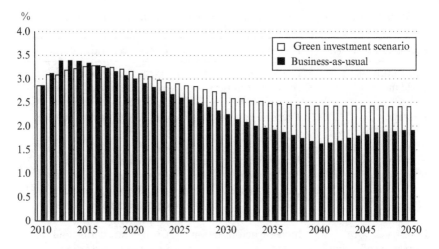

Source: UNEP, 'Towards a Green Economy: Pathways to Sustainable Development and Poverty Eradication', Synthesis Report (UNEP 2011) 1–2, available at <www.unep. org/greeneconomy/Portals/88/documents/ger/GER_synthesis_en.pdf> (last accessed on 1 August 2015).

Figure 6.2   Comparison of GDP growth in a green investment and a BAU scenario

The results are then compared across different indicators that capture economic development and environmental improvements.

As shown in Figure 6.2, in the Green Economy scenario (with light bars), growth in global GDP, is projected to be higher than in the business-as-usual scenario (with dark bars) after ten years, while in the short-term the BAU still yields better results. When looking at some of the sectors in more detail, in fisheries – where major stocks are already collapsing – a reduction of capacity is required in the short-term until fishing can resume at a sustainable level. In other sectors, e.g. for investments in energy efficiency in buildings or public transport, growth in income and jobs are more immediate. In the medium and longer term most of the sectors become competitive vis-à-vis their respective BAU scenario.

But the higher GDP growth in the BAU scenario, which continues over the first years of the projection, comes at a high price. When taking into account environmental indicators, the BAU scenario is characterized by: (i) continued high-carbon intensity with associated environmental

---

and continued unsustainable exploitation of natural resources). See UNEP, 'Towards a Green Economy' (n 29) 515.

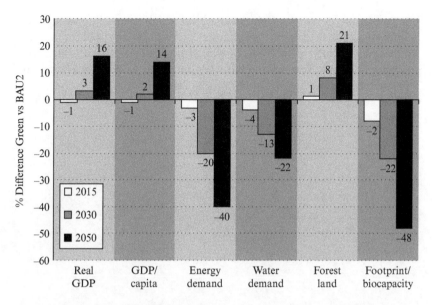

Source: UNEP, 'Towards a Green Economy: Pathways to Sustainable Development and Poverty Eradication', Synthesis Report (UNEP 2011) 1–2, available at <www.unep. org/greeneconomy/Portals/88/documents/ger/GER_synthesis_en.pdf> (last accessed on 1 August 2015).

Figure 6.3   Impacts of the green investment scenario relative to BAU for selected variables (percentage change)

impacts, especially in terms of the long-term concentration of atmospheric greenhouse gases (GHG), resulting in temperature increases most likely around 4° Celsius; and (ii) the further depletion of natural capital and a global ecological footprint more than two times the available bio-capacity of the Earth by 2050.

In the Green Economy scenario, while economic growth can be sustained over 40 years, natural resources are not depleted and emissions are not increasing at the same rate as in the BAU scenario as shown in Figure 6.3 above. There is also an indication that natural capital would be built up again through the green investments, which will help to secure the livelihoods of many poor people depending on them.

In the simulations, an annual investment of 2 percent of global GDP up to 2050, can potentially double fish stocks, increase forest land by one-fifth, and lower global demand for water by about 20 percent, as compared to BAU. It is also projected to reduce energy-related CO_2 emissions by about one-third by 2050 compared to current levels, and bring the atmospheric concentration of emissions closer to a level which keeps climate change in a limit of a global temperature change of 2° Celsius. By maintaining and building up natural capital and in turn mitigating resource scarcity, these investments would provide the basis for enhanced human well-being, and sustained economic growth over the next 20 to 40 years.[76]

---

[76] Ibid., 509 ff.

7. Resource recovery from electric and electronic waste

Mathias Schluep[1]

EXECUTIVE SUMMARY

The increasing penetration of society with electrical and electronic equipment (EEE) is resulting in growing volumes of waste. Typical of this waste is the combination of its intrinsic value due to the high content of basic and precious metals, with health and environmental hazards caused by the occurrence of toxic substances in combination with inadequate recycling practices. Based on the principle of extended producer responsibility (EPR), industrialized countries have legislated waste electrical and electronic equipment (WEEE) management. As a consequence, take-back schemes have been established and innovative recycling technologies developed to recover resources from this waste stream. Although collection rates are often low and technical as well as operational aspects to recover scarce and critical metals still need to be addressed, developing countries are catching up with both increasing waste volumes and addressing the challenge with legislation and policies. Inefficient and harmful recycling technologies in the informal sector are, however, still prevailing.

7.1 INTRODUCTION

The use of Electrical and Electronic Equipment (EEE) has grown rapidly in recent decades. Expanded functionalities and decreasing prices have influenced consumer behaviour. As a consequence, Waste Electrical and Electronic Equipment (WEEE or e-waste) has become the fastest-growing waste stream worldwide.[2] Early experiences in industrialized countries

---

[1] The views expressed herein are the views of the author and do not necessarily reflect the views of the World Resource Forum Association.

[2] M. Schluep, C. Hagelueken, R. Kuehr, F. Magalini, C. Maurer, C: Meskers, E. Mueller, and F. Wang, 'Recycling - from e-waste to resources: Sustainable innovation

have shown that municipalities were not adequately equipped or staffed to handle a complex waste stream such as WEEE. As a result of mainly two new paradigms, a change began in the management of e-waste: the 'closed loop economy' and the 'extended producer responsibility' (EPR). Based on EPR, producers initiated take-back schemes, either individually or collectively, as a group of producers or as members of national producer responsibility organizations (PRO), to manage the financing of WEEE flows and related processing steps. National authorities started to address this concept in their waste regulations. The recycling industry went through a rapid evolution where specialists emerged, among others, for manual dismantling, mechanical processing or final refining of secondary raw materials. Innovative technologies to separate hazardous components and recover resources from waste emerged.

In developing countries and emerging economies, these concepts were not adapted until recently. Reports from NGOs on the environmental and health issues related to poor WEEE management as well as various international cooperation projects addressing those challenges[3] increased the priority of WEEE management among the environmental issues requiring special legislative attention. As a result, WEEE legislation based on EPR has been established in a number of developing countries in recent years. Summaries of rapid developments can be found as global overviews,[4] or in publications focusing on specific regions: Africa,[5] Asia,[6] and Latin America.[7] Since 2011, a few African (i.e., Ghana and Kenya) and Latin

---

and technology transfer industrial sector studies' (UNEP 2009) available at <www.ewasteguide.info/files/UNEP_2009_eW2R.PDF> (last accessed on 7 August 2015).

[3]   M. Schluep, E. Müller, L. M. Hilty, D. Ott, R. Widmer, and H. Böni, 'Insights from a decade of development cooperation in e-waste management' in *Proceedings of the First International Conference on Information and Communication Technologies for Sustainability* (ETH Zurich, 2013) available at <www.ewasteguide.info/files/Schluep_2013_ICT4S.pdf> (last accessed on 7 August 2015).

[4]   F. O. Ongondo, I. D. Williams, and T. J. Cherrett, 'How are WEEE doing? A global review of the management of electrical and electronic wastes' (2011) 31(4) *Waste Management* 714; V. Goodship and A. Stevels, *Waste Electrical and Electronic Equipment (WEEE) Handbook* (Woodhead Publishing Ltd 2012).

[5]   Secretariat of the Basel Convention, 'Where are WEee in Africa? Findings from the Basel Convention e-Waste Africa Programme' (2011) available at <www.basel.int/Implementation/Ewaste/EwasteinAfrica/Overview/PublicationsReports/tabid/2553/Default.aspx> (last accessed on 7 August 2015); M. Schluep, 'WEEE management in Africa' in Goodship and Stevels (n 3).

[6]   X. Chi, M. Streicher-Porte, M. Y. L. Wang, and M. A. Reuter, 'Informal electronic waste recycling: A sector review with special focus on China' (2011) 31(4) *Waste Management* 731.

[7]   D. Garcés and U. Silva, 'Guía de contenidos legales para la gestión de los

American countries (i.e., Colombia and Peru) have introduced EPR as the core principle in their national WEEE legislation. Kenya, for example, published draft WEEE regulations in 2013 for public consultation. In Peru, the 'Reglamento de Gestión y Manejo de Residuos Eléctricos y Electrónicos – RAEE' introduced EPR as a new principle in national waste legislation in 2012.[8] However, with most challenges to implement such policies still in the future, recycling industries are developing slowly and still in an uncontrolled manner. Harmful technologies to recover the more easily accessible waste fractions prevail. This chapter intends to present currently applied methods and technologies to recycle e-waste not only from an OECD perspective but also from a developing world view. It furthermore intends to discuss the most important trends and our preparedness to address related challenges from a technology point of view.

7.2   VOLUMES AND COMPOSITION OF WEEE

WEEE or e-Waste is often misunderstood as comprising only computers and related IT equipment. But, according to the OECD, e-waste is 'any appliance using an electric power supply that has reached its end-of-life'. In this chapter, WEEE and e-waste are used as synonyms, and include all the ten categories according to the European Union (EU) WEEE Directive.[9]

7.2.1   Volumes of ICT Waste

Usage of EEE in a society and the corresponding stocks and flows are important elements for the design of management systems, both from a waste and from a material management perspective.

For industrialized countries recent statistics estimate the worldwide quantity of EEE put on the market in 2012 at roughly 65 million tons and the corresponding generation of WEEE at almost 49 million tons.[10]

---

residuos electrónicos', Centro de Derecho Ambiental, Facultad de Derecho, Universidad de Chile (2010).

[8]   Ministerio de Medio Ambiente (MINAM) del Perú, Decreto Supremo N° 001-2012-MINAM: Reglamento nacional para la gestión y manejo de los residuos de aparatos eléctricos y electrónicos (2012).

[9]   European Union, Directive 2012/19/EU of the European Parliament and of the Council of 4 July 2012 on waste electrical and electronic equipment (WEEE) (2012) OJ L 197, 38–71.

[10]   C. P. Baldé, F. Wang, R. Kuehr and J. Huisman, 'The Global E-Waste Monitor – 2014' (Bonn, United Nations University 2015) available at <http://i. unu.

Table 7.1 below presents reported data from selected European countries. WEEE collected varies greatly between e.g., 1.2 kg per capita in Romania and 22.0 kg per capita in Norway. The large difference is due to lower EEE penetration and less developed WEEE take-back systems in Romania in comparison to Norway.

For developing countries there is only limited data availability; the quantification of WEEE volumes in developing countries is an iterative process, often based on a combined top-down and bottom-up approach. Figures on imports of EEE can often be derived from statistical data, while consumer stocks and disposal volumes need to be assessed through surveys. Informal waste collection is least documented, for which reason WEEE quantities are often assessed by assigning lifetimes to specific products. Through additional field investigations as well as

Table 7.1   EEE put on the market and WEEE collected in 15 selected European countries in 2010 [11]

Country	Population (millions)	EEE put on the market (t)	WEEE collected (t)	WEEE collected per capita (kg/person)
Austria	8.4	165,810	74,256	8.8
Belgium	10.8	294,530	105,557	9.8
Denmark	5.5	147,557	82,931	15.1
Finland	5.4	148,157	50,867	9.4
France	65.4	1,635,493	433,959	6.6
Germany	81.8	1,730,794	777,035	9.5
Greece	11.3	178,260	46,528	4.1
Italy	60.3	1,117,406	268,216	4.4
Latvia	2.2	15,290	4,288	1.9
Norway	4.9	181,579	107,767	22.0
Portugal	10.6	157,065	46,673	4.4
Romania	21.5	151,317	26,247	1.2
Spain	47.0	746,801	158,100	3.4
Sweden	9.4	232,403	161,444	17.2
United Kingdom	62.0	1,534,576	479,356	7.7

edu/media/ias.unu.edu-en/news/7916/Global-E-waste-Monitor-2014-small.pdf> (last accessed on 11 August 2015).

[11]   European Commission, 'Environmental Data Centre on Waste; Key waste streams; Waste electrical and electronic equipment (WEEE)' (2014) available at <http://epp.eurostat.ec.europa.eu/portal/page/portal/waste/key_waste_streams/waste_electrical_electronic_equipment_weee> (last accessed on 7 August 2015).

Table 7.2   *PCs put on the market and estimated PC waste generation in selected developing countries according to various country assessments*

Country	Assess. Year	Pop (mill)	PCs put on the market (t)	PC waste generated (t)	PC waste generated per capita (kg)
Ghana[1]	2009	24.3	16,650	6,400	0.3
Kenya[2]	2007	40.9	5,200	440	0.01
South Africa[3]	2007	50.0	32,000	19,400	0.4
Uganda[4]	2007	33.8	700	1,300	0.2
China[5]	2007	1,339.2	419,100	300,000	0.2
India[6]	2007	1,184.7	140,800	56,300	0.01
Brazil[7]	2005	193.4	no data	96,800	0.5
Chile[8]	2010	17.1	12,600	5,300	0.3
Colombia[9]	2006	45.6	13,600	6,500	0.1
Peru[10]	2006	29.5	7,000	6,000	0.2

Notes:
1   Y. Amoyaw-Osei, O. O. Agyekum, J. A. Pwamang, E. Mueller, R. Fasko, and M. Schluep, 'Ghana e-waste country assessment' Green Advocacy, Ghana & Empa, Switzerland (2011) available at <www.ewasteguide.info/files/Amoyaw-Osei_2011_GreenAd-Empa.pdf> (last accessed on 7 August 2015).
2   Schluep et al 'Recycling' (n 2); T. Waema and M. Mureithi, 'E-waste management in Kenya' Kenya ICT Action Network (KICTANet) (2008) available at <http://ewasteguide.info/files/Waema_2008_KICTANet.pdf> (last accessed on 7 August 2015).
3   Schluep et al 'Recycling' (n 2); A. Finlay and D. Liechti, 'E-Waste assessment South Africa' Openresearch, Empa (2008) available at <http://ewasteguide.info/files/Finlay_2008_eWASA.pdf> (last accessed on 7 August 2015).
4   Schluep et al 'Recycling' (n 2); J. Wasswa and M. Schluep, 'E-waste assessment in Uganda: A situational analysis of e-waste management and generation with special emphasis on personal computers', Uganda Cleaner Production Center, Empa (2008) available at <http://ewasteguide.info/files/Wasswa_2008_UCPC-Empa.pdf> (last accessed on 7 August 2015).
5   Schluep et al 'Recycling' (n 2).
6   Ibid.
7   Ibid.; G. Rocha, 'Diagnosis of waste electric and electronic equipment generation in the State of Minas Gerais', Fundacao Estadual do Meio Ambiente (FEAM), Governo Minas (2009) available at <http://ewasteguide.info/files/Rocha_2009_en.pdf> (last accessed on 7 August 2015).
8   B. Steubing, H. Böni, M. Schluep, U. Silva, and C. Ludwig, 'Assessing computer waste generation in Chile using material flow analysis' (2010) 30 *Waste Management* 473; B. Steubing, 'E-Waste generation in Chile, situation analysis and estimation of actual and future computer waste quantities using material flow analysis', Master's thesis, Ecole Polytechnique Federal de Lausanne (2007).
9   Schluep et al 'Recycling' (n 2); D. Ott, 'Gestión de Residuos Electrónicos en Colombia: Diagnóstico de Computadores y Teléfonos Celulares', Informe Final, Swiss Federal Laboratories for Materials Testing and Research (Empa), Centro Nacional de

Table 7.2 (continued)

Producción Mas Limpia (CNPMLTA) (2008) available at <http://ewasteguide.info/files/Ott_2008_Empa-CNPMLTA.pdf> (last accessed on 7 August 2015).

10 Schluep et al 'Recycling' (n 2); O. Espinoza, L. Villar, T. Postigo, and H. Villaverde, 'Diagnóstico del manejo de los residuos electrónicos en el Perú' Institute for the Development of Social Economy (IPES), Swiss Federal Laboratories for Materials Testing and Research (Empa) (2008) available at <http://ewasteguide.info/files/Espinoza_2008_IPES-Empa.pdf> (last accessed on 7 August 2015).

---

interviews, meetings, and workshops with stakeholders, valuable information such as transfer coefficients between processes, downstream processes of materials, and information about material quality can be obtained.

Various WEEE assessments performed between 2005 and 2012 have revealed figures on Personal Computer (PC) imports and PC waste as shown in Table 7.2. The data on PC waste are indicative and are derived from material flow assessments.

7.2.2 Composition of WEEE

The perception of WEEE has developed over the years from a waste problem, which can cause environmental damage and health issues, to an opportunity: ICT components, for example, contain a variety of metals for which recovery is economically attractive (Table 7.3). The metal concentrations often exceed the concentrations found in natural ores.[12] The Kloof gold mine in South Africa, for instance, has gold concentrations of approximately 6 ppm gold,[13] whereas in mobile phones this concentration can be up to 100 times higher. Similar situations can be found when comparing silver and palladium concentrations in natural ores with concentrations in ICT components.

Compared to annual production volumes, the demand for metals used in EEE reaches significant levels (Table 7.4). This highlights the relevance of WEEE as a secondary resource. Consequently, inefficient treatment of

---

[12] C. Hagelueken, 'Towards bridging the material loop. How producers and recyclers can work together', presented at the EU–US Workshop on 'Mineral Raw Material Flows and Data', Brussels, 13 September 2012, available at <https://ec.europa.eu/eip/raw-materials/en/system/files/ged/27%20three_hagelueken_eu_us_2012_09_en.pdf> (last accessed on 11 August 2015).

[13] Kloof Gold Mines, 'Mineral resources and mineral reserves overview' (2009) available at <www.goldfields.co.za/reports/rr_2009/tech_kloof.php> (last accessed on 4 May 2014).

Table 7.3   Content of gold (Au), silver (Ag) and palladium (Pd) in ICT devices [14]

Device	Au		Ag		Pd	
	(mg)	(ppm)	(mg)	(ppm)	(mg)	(ppm)
PC	316–338	21–23	804–2,127	54–142	146–212	10–14
Laptop	246–250	85–86	440	152	50–80	17–28
Tablet	131	215	26	43	no data	no data
Mobile phone	50–69	455–627	127–715	1,155–6,500	10–37	91–336

WEEE may lead to a systematic loss of secondary materials.[15] Hence, the appropriate handling of WEEE both prevents environmental and health issues and contributes to more sustainable use of raw materials.

WEEE also contains toxic and hazardous substances, for example, heavy metals such as mercury, cadmium, lead, and chromium, or persistent organic pollutants (POPs), which can be found in plastic casings or in printed wiring boards (PWB).[16] Some of these substances have been regulated, and their use has been restricted for new equipment through the EU RoHS Directive.[17] Other substances have been banned, but are still allowed for certain applications (for instance, mercury in energy-saving lamps) or are still present in older equipment. WEEE and its components may therefore pose a significant health risk not only due to their primary constituents, but also as a result of improper management of by-products either used in the recycling process (such as cyanide for leaching gold) or generated by chemical reactions (such as dioxins through the burning

[14]   K. Sander et al, 'Abfallwirtschaftliche Produkteverantwortung unter Ressourcenaspekten Projekt RePRO, Meilensteinbericht August 2012', Bundesministerium für Umwelt, Naturschutz und Reaktorsicherheit, Meilensteinbericht FKZ 371195318 (2012) available at <www.oekopol.de/archiv/material/603_RePro_Meilensteinbericht_1.pdf> (last accessed on 11 August 2015).

[15]   Schluep et al 'Recycling' (n 2).

[16]   P. A. Wäger, M. Schluep, E. Müller, and R. Gloor, 'RoHS regulated substances in mixed plastics from waste electrical and electronic equipment' (2012) 46(2) *Environmental Science & Technology* 628.

[17]   European Union, Directive 2011/65/EU of the European Parliament and of the Council of 8 June 2011 on the restriction of the use of certain hazardous substances in electrical and electronic equipment (recast) (2011) OJ L 174, 88–110.

Table 7.4   Important metals used for EEE[18]

Metal	Primary production* (t/y)	By-product of	Demand for EEE (t/y)	Demand/ production (%)	Main applications
Ag	20,000	Pb, Zn	6,000	30	Contacts, switches, solders . . .
Au	2,500	(Cu)	300	12	Bonding wire, contacts, integrated circuits . . .
Pd	230	PGM	33	14	Multilayer capacitors, connectors
Pt	210	PGM	13	6	Hard disks, thermocouples, fuel cells
Ru	32	PGM	27	84	Hard disks, plasma displays
Cu	15,000,000		4,500,000	30	Cables, wires, connectors . . .
Sn	275,000		90,000	33	Solders
Sb	130,000		65,000	50	Flame retardants, CRT glass
Co	58,000	Ni, Cu	11,000	19	Rechargeable batteries
Bi	5,600	Pb, W, Zn	900	16	Solders, capacitors, heat sinks . . .
Se	1,400	Cu	240	17	Electro-optic devices, copiers, solar cells
In	480	Zn, Pb	380	79	LCD glass, solders, semiconductors
Total			4,670,000		

Notes:
* Based on demand in 2006.
Acronyms: PGM = Platinum Group Metals; CRT = Cathode Ray Tube; LCD = Liquid Crystal Display.

of cables). Due to its properties, WEEE is generally considered to be hazardous waste under the Basel Convention.

7.3   THE RECYCLING CHAIN

The recycling chain for WEEE consists of three main subsequent steps: (i) collection; (ii) pre-processing (including sorting, dismantling, mechanical

---

[18]   Ibid.

treatment); and (iii) end-processing (including refining and disposal).[19] Usually, for each of these steps specialized operators and facilities exist. The material recovery efficiency of the entire recycling chain depends on the efficiency of each step and on how well the interfaces between these interdependent steps are managed. If, for example, for a certain material the efficiency of collection is 50 per cent, the combined pre-processing efficiency is 70 per cent and the refining (materials recovery) efficiency 95 per cent, the resulting net material yield along the chain would be only 33 per cent.

Concepts and processes applied in the recycling chain can vary considerably from each other in different regions and countries with individual strengths and weaknesses. The main differences can be found between OECD countries with a prevailing formal sector and developing countries with a dominant informal sector. Figure 7.1 compares the recycling efficiency between a common formal system in Europe and the informal sector in India for the overall gold yield from PWBs. While both scenarios indicate similar (low) overall metal recovery efficiencies, both have their weaknesses and strengths in different steps of their respective recycling chain. An analysis of strengths, weaknesses, opportunities and threats (SWOT analysis) for a formal versus an informal system is summarized in Table 7.5 below.

7.4   FORMAL RECYCLING

7.4.1   Collection

In formal WEEE schemes, municipal collection points and/or retailers' take-back obligations are the backbone of collection. In the European Union EU, take-back obligations in the previous WEEE Directive entailed only municipal collection points. As in some countries take-back quantities are still rather low, the recast of the WEEE Directive[20] has defined responsibilities for distributors to take back equivalent equipment they sell on a one-to-one basis (an obligation for the customer to buy equipment with an equivalent function) and free of charge. In 2010, the quantities collected in the EU ranged between 1.1 (Romania) and 15.9 kg/capita (Sweden), with ten countries still not reaching the required 4.0 kg/

---

[19]   Schluep et al 'Recycling' (n 2).
[20]   See EU Directive 2012/19/EU (n 9).

System	Collection	Pre-processing	End-processimg	Net Yield
Formal e.g. Europe[21]	60% formal take - back system	25% mainly mechanical processes	95% integrated smelter	15%
Informal e.g. India[22]	80% individual collectors	50% manual sorting and dismantling	50% backyard leaching	20%

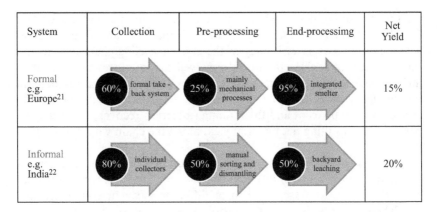

Figure 7.1   Recycling efficiency between a common formal system in Europe and the informal sector in India for the overall gold yield out of printed wiring boards[23]

capita minimum collection quantity. From 2016 onward, a minimum collection rate of 45 per cent of EEE placed on the market in the three preceding years has to be achieved by each Member State.

---

[21]   J. Huisman, F. Magalini, R. Kuehr, C. Maurer, S. Ogilvie, J. Poll, C. Delgado, E. Artim, J. Szlezak, and A. Stevels, '2008 Review of Directive 2002/96 on Waste Electrical and Electronic Equipment (WEEE), Final Report' (United Nations University 2008) available at <http://ec.europa.eu/environment/waste/weee/pdf/final_rep_unu.pdf> (last accessed on 7 August 2015); P. Chancerel, C. E. M. Meskers, C. Hagelüken, and V. S. Rotter, 'Assessment of Precious Metal Flows During Preprocessing of Waste Electrical and Electronic Equipment' (2009) 13(5) *Journal of Industrial Ecology* 791.

[22]   M. Keller, 'Assessment of gold recovery processes in Bangalore, India and evaluation of an alternative recycling path for printed wiring boards', Master's thesis, Swiss Federal Institute of Technology (ETH), Materials Science and Technology Research Institute (Empa) (2006) available at <www.empa.ch/plugin/template/empa/*/59244> (last accessed on 7 August 2015); D. Rochat, C. Hagelüken, M. Keller, and R. Widmer, 'Optimal Recycling for Printed Wiring Boards (PWBs) in India', in R'07 Recovery of Materials and Energy for Resource Efficiency (2007) 12, available at <http://ewasteguide.info/files/Rochat_2007_R07.pdf> (last accessed on 7 August 2015).

[23]   UNEP, 'Metal Recycling – Opportunities, Limits, Infrastructure' Draft No. 3 (Paris, UNEP 2012) available at <www.unep.org/resourcepanel/Portals/24102/PDFs/Metal_Recycling_Full_Report.pdf> (last accessed on 7 August 2015).

Table 7.5   SWOT analysis of the e-waste recycling chain in formal vs informal scenarios

Strengths	Access to state-of-the-art end-processing facilities with high metal recovery efficiency[1]	High collection efficiency[2] Efficient deep manual dismantling and sorting[3]; Low labour costs give advantage of manual techniques over mechanical technologies in the pre-processing steps[4]
Weaknesses	Low efficiency in collection[5]; Often low efficiency in (mechanized) pre-processing steps[6]	Medium efficiency in dismantling and sorting[7]; Low efficiency in end-processing steps[8] coupled with adverse impacts on humans and the environment[9]
Opportunities	Improvement of collection efficiency; Technology improvement in pre-processing steps	Improvement of efficiency in the pre-processing steps through skills development for dismantling and sorting;[10] Implementation of alternative business models, providing an interface between informal and formal sector[11]
Threats	'Informal' activities in the collection systems	Bad business practice (bribery, cherry picking of valuables only, illegal dumping of non-valuables, etc.); Lacking government support (no acceptance of informal sector, administrative hurdles for receiving export licenses, etc.)

Notes:
1   Schluep et al 'Recycling' (n 2).
2   Schluep et al 'Recycling' (n 2).
3   S. Gmuender, 'Recycling – from waste to resource: assessment of optimal manual dismantling depth of a desktop PC in China based on eco-efficiency calculations', Master's thesis, Swiss Federal Institute of Technology (EPFL) / Swiss Federal Laboratories for Materials Testing and Research (Empa) (2007) available at <http://ewasteguide.info/files/Gmuender_2007_ETHZ-EMPA.pdf> (last accessed on 7 August 2015).
4   Schluep et al 'Recycling' (n 2).
5   Huisman et al (n 30).
6   Chancerel et al (n 30).
7   Keller (n 31).
8   Ibid.
9   F. Wang, J. Huisman, C. Meskers, M. Schluep, A. Stevels, and C. Hagelueken, 'The Best-of-2-Worlds philosophy: Developing local dismantling and global infrastructure network for sustainable e-waste treatment in emerging economies' (2012) 32 *Waste Management* 2134.
10   D. Rochat, W. Rodrigues, and A. Gantenbein, 'India: Including the existing informal sector in a clean e-waste channel', in Proceedings of the 19th Waste Management Conference (WasteCon2008) (2008) available at <http://ewasteguide.info/files/Rochat_2008_WasteCon.pdf> (last accessed on 7 August 2015).
11   Ibid.

7.4.2   Pre- and End-processing

Selective treatment of components like printed circuit boards, capacitors, batteries, mercury containing components and others, in most cases require initial manual dismantling steps. In countries with low collection quantities and low labour costs, manual dismantling is a preferred option, since mechanical treatment is not economically viable. In other countries, pre-treatment is a combination of manual and mechanical process steps. Mechanical treatment includes crushing and separation of different metal- or plastic-rich fractions or mixtures of the two, which then undergo further segregation steps such as conductivity (eddy-current) or density separation (swim sink). End treatment of metal fractions is a combination of different wet-chemical and metallurgical processes with the aim of obtaining pure fractions that can become secondary commodities. Plastic fractions are separated into those suitable for material recycling and others that have to be finally disposed of in incineration plants or landfills.

7.4.3   Health Hazards and Environmental Impacts

Formal recycling processes have the potential to endanger health and the environment. Direct impacts on health are caused by dust in indoor air generated during dismantling and mechanical treatment processes (e.g., from plastic shredding or treatment of CRT) or non-conformities with occupational health requirements. Indirect impacts on human health may be caused by air pollution related to incineration processes that are not equipped with adequate gas purification systems and dust retention.

Mixed plastics fractions from WEEE still contain regulated brominated flame retardants (BFRs). High average concentrations of BFRs mainly originate from small household appliances for high temperature applications, CRT monitors, and consumer equipment, in particular CRT TVs.[24]

Primary production of resources, i.e., mining, concentrating, smelting, and refining, is energy-intensive and hence has a significant carbon dioxide (CO_2) impact. 'Mining' of old computers to recover the materials they contain – if done in an environmentally sound manner – needs only a fraction of this energy input.[25]

---

[24]   Wäger et al (n 27).
[25]   S. Grimes, J. Donaldson and G. C. Gomez, 'Report on the environmental benefits of recycling', Bureau of International Recycling (BIR) (2008) available at <https://cari-acir.org/wp-content/uploads/2014/08/BIR_CO2_report.pdf> (last accessed on 7 August 2015).

7.5   INFORMAL RECYCLING

7.5.1   The Informal Sector

In developing countries, waste management is mostly performed by a large urban workforce, usually referred to as the 'informal sector', making a living by collecting, sorting, recycling, and selling valuable materials recovered from waste.[26] The marginalized poor account for the majority of the informal sector. They often include groups from ethnic or religious minorities or rural migrants. Women and children constitute a significant proportion of the workforce, operating either illegally or in a legal grey zone and with different levels of organization.[27]

Even though informality has been the subject of political and scientific discussions for decades,[28] there is no clear definition of the informal sector. Yet all definitions point toward similar elements and patterns[29] generally described by the International Labor Organization (ILO) as including all economic units that are not regulated by the state and all economically active persons who do not receive social protection through their work.[30]

Collection, manual dismantling, open burning to recover metals, and open dumping of residual fractions are the usual practice in most countries. In smaller and less developed economies, these activities are usually performed by individuals, as volumes are too small to trigger the informal sector to specialize in WEEE recycling on a larger scale. Larger

---

[26]   M. Medina, 'Informal recycling and collection of solid wastes in developing countries: Issues and opportunities', United Nations University/Institute of Advanced Studies, Tokyo, Japan, UNU/IAS Working Paper No. 24 (1997) available at <www.gdrc.org/uem/waste/swm-ias.pdf> (last accessed on 7 August 2015).

[27]   K. Lundgren, 'The global impact of e-waste – Addressing the challenge', International Labour Office, Programme on Safety and Health at Work and the Environment (SafeWork), Sectoral Activities Department (SECTOR) (2012) available at <www.ilo.org/wcmsp5/groups/public/@ed_dialogue/@sector/documents/publication/wcms_196105.pdf> (last accessed on 7 August 2015).

[28]   X. Chi, M. Streicher-Porte, M. Y. L. Wang and M. A. Reuter, 'Informal electronic waste recycling: A sector review with special focus on China' (2011) 31(4) *Waste Management* 731.

[29]   M. Schluep, 'Informal waste recycling in developing countries' in E. Worrell and M. A. Reuter (eds), *Handbook of Recycling: State-of-the-Art for Practitioners, Analysts, and Scientists* (Elsevier 2014).

[30]   ILO, 'Men and women in the informal economy: A statistical picture' International Labour Office, (2002) available at <www.ilo.org/wcmsp5/groups/public/---dgreports/---stat/documents/publication/wcms_234413.pdf> (last accessed on 7 August 2015).

economies, especially countries in transition such as India and China,[31] as well as countries subject to intense trade in second-hand equipment and illegal waste shipments, such as Ghana and Nigeria,[32] display a substantial organized informal sector.

7.5.2 Collection

In contrast to formalized take-back schemes where consumers (indirectly) pay for collection and recycling, in developing countries it is usually the waste collectors who pay consumers for obtaining their obsolete appliances and scrap material.[33] As a result, the informal waste sector is often organized in a network of individuals and small businesses of collectors, traders, and recyclers, each adding value and creating jobs at every point in the recycling chain.[34] Since the valuable components of the products collected usually generate an income higher than the price to be paid to get the product, the informal waste sector achieves collection rates of up to 95 per cent of waste generated,[35] which is far above what can be achieved by today's formalized take-back schemes.[36]

7.5.3 Pre- and End-processing

As labour costs are low in developing countries and countries in transition, and because of the lack of access to know-how and technology, informal and formal recyclers apply labour-intensive pre-processing technologies such as manual dismantling to separate the heterogeneous materials and components. A comparative study of pre-processing scenarios revealed that material recovery efficiency improves with the intensification of manual dismantling.[37] Hence, manual recycling practices in developing

---

[31] Chi et al (n 49); A. Sepúlveda, M. Schluep, F. G. Renaud, M. Streicher, R. Kuehr, C. Hagelüken, and A. C. Gerecke, 'A review of the environmental fate and effects of hazardous substances released from electrical and electronic equipments during recycling: Examples from China and India' (2010) 30 *Environmental Impact Assessment Review* 28; Wang et al (n 42).

[32] Secretariat of the Basel Convention, 'Where are WEee in Africa?' (n 5).

[33] UNEP, 'Metal Recycling' (n 32).

[34] D. Sinha-Khetriwal, P. Kraeuchi, and M. Schwaninger, 'A comparison of electronic waste recycling in Switzerland and in India' (2005) 25 *Environmental Impact Assessment Review* 492.

[35] Secretariat of the Basel Convention, 'Where are WEee in Africa?' (n 5).

[36] Huisman et al (n 30).

[37] Chancerel et al (n 30); Wang et al (n 41).

countries do display advantages, such as low investment costs, creation of jobs, and high material recovery efficiency.[38]

Subsequent to manual pre-processing, further 'refining' techniques, such as de-soldering of PWBs and subsequent leaching of gold, silver, and palladium, have been observed especially in the informal sectors in India and China.[39] A pilot project in Bangalore, India, demonstrated that besides being hazardous, informal end-processing or refining practices also have poor recovery efficiency. Improper sorting of PWBs and subsequent wet chemical leaching processes for the recovery of gold, for example, revealed a combined yield of only 25 per cent.[40] In contrast, today's state-of-the-art integrated smelters, as used in most formalized recycling systems, achieve gold recovery efficiencies as high as 95 per cent.[41]

7.5.4 Health Hazards and Environmental Impacts

Informal WEEE management often fills the void created by the absence of a legal framework as well as the lack of capacity and resources for a formal waste collection and treatment system.

Due to their daily contact with garbage, people working in informal waste management are exposed to various health threats, including injuries, diseases, and both acute and chronic health effects. Serious health effects and impacts on the environment are likely especially for workers processing waste streams containing hazardous substances, such as WEEE.[42] Emissions stem from: (i) hazardous substances which are constituents of the waste; (ii) auxiliary substances used in recycling techniques; and (iii) by-products formed by the transformation of primary constituents. The activities of WEEE recycling in the informal sector involve sorting as well as separation with the final aim of extracting valuable materials such as copper, gold, silver, and other base and precious metals. The processes applied in the exploitation of metals are of particular concern since they

---

[38] Schluep et al 'Recycling' (n 2).

[39] Sepúlveda et al (n 52).

[40] Keller (n 31); Rochat et al (n 31).

[41] P. Chancerel and V. Rotter, 'Stop wasting gold – How a better mining of end-of-life electronic products would save precious resources', in *R'09 World Congress* (2009) available at <www.ewasteguide.info/files/Chancerel_2009_R09. pdf >(last accessed on 7 August 2015).

[42] Sepúlveda et al (n 52); K. Grant, F. C. Goldizen, P. D. Sly, M.-N. Brune, M. Neira, M. van den Berg, and R. E. Norman, 'Health consequences of exposure to e-waste: a systematic review', *Lancet Global Health* (Oct. 2013) available at <http:// eprints.qut.edu.au/79831/1/Health%20consequences%20of%20exposure%20to%20 ewaste.pdf> (last accessed on 7 August 2015).

cause a variety of health and environmental hazards. A literature review concerning emissions caused by informal recycling activities has shown high concentrations of lead, polybrominated diphenyl ethers (PBDE), dioxins, and furans in all environmental pathways (soil, air, water, bottom ash, and river sediments).[43]

Practices for recovering metals such as copper, iron, and aluminium through burning of cables containing PVC insulation have been identified as a major source of dioxin.[44] Dioxin emissions from cable burning, for instance in the greater Accra region in Ghana alone, are estimated to correspond to about 0.3 per cent of total dioxin emissions in Europe.[45] In China and India, a review of various studies underlined very high levels of dioxin in air, bottom ash, dust, soil, water, and sediments in informal recycling areas, which sometimes exceeded the reference values generally observed in urban areas by several orders of magnitude.[46]

Recent measurements in Accra, Ghana, also indicate increasing levels of BFRs in breast milk, which are associated with the informal recycling of WEEE.[47]

BFRs contained in mixed plastics from WEEE are substances of concern due to the existing practices of plastic recycling in developing countries and the potential risk of cross-contaminating secondary plastics in applications where BFRs are not required or banned. A recent sampling campaign in the informal plastic recycling sector in Delhi, India, confirmed that secondary plastic is often contaminated with BFRs. This indicates that mixing of plastics from WEEE with additive-free plastics from other waste types does occur.[48]

7.5.5 Socio-economic Impacts

Safety- and health-related impacts were observed in many developing countries, leading to direct effects on the workers and the local communi-

---

[43] Sepúlveda et al (n 52).

[44] Secretariat of the Basel Convention, 'Where are WEee in Africa?' (n 5).

[45] Amoyaw-Osei et al (n 12).

[46] Sepúlveda et al (n 52).

[47] K. A. Asante, S. Adu-Kumi, K. Nakahiro, S. Takahashi, T. Isobe, A. Sudaryanto, G. Devanathan, E. Clarke, O. D. Ansa-Asare, S. Dapaah-Siakwan, and S. Tanabe, 'Human exposure to PCBs, PBDEs and HBCDs in Ghana: Temporal variation, sources of exposure and estimation of daily intakes by infants', (2011) 37(5) *Environment International* 921.

[48] Toxics Link and Empa, 'Improving plastics management in Dehli. A report on WEEE plastics recycling' (2012) available at <www.ewasteguide.info/files/ToxicsLink_2012_PlasticRecycling.pdf> (last accessed on 7 August 2015).

ties as outlined in the previous section. As most of the workforce belongs to the informal sector, WEEE recycling does not feature formalized workers' participation mechanisms which results in the lack of worker rights.

In Ghana, child labour was observed for cable-burning activities and for manual dismantling practices such as breaking CRT monitors. Using stones, hammers, heavy metal rods, and chisels to recover copper, steel, and plastic casings from CRT often results in the workers inhaling hazardous cadmium dust and other pollutants.[49]

Income levels vary depending on the profit which can be generated by selling the obsolete equipment to recyclers in relation to the price paid for acquiring the equipment. In Ghana, a collector can earn 70–140 USD per month, whereas recyclers can earn 175–285 USD a month. In Nigeria, the corresponding figures are 67–100 USD per month for collectors and recyclers. However, these figures are based on calculated incomes based on business profits and do not consider indirect costs and externalities.

In Pakistan, children aged between six and 18 search for valuable materials in potentially toxic ash. They work in all stages of the chain, from collecting and dismantling equipment to burning wires and motherboards, separating metals, melting solders, and acid processes.[50]

The ILO states that the existing ILO conventions are intended to address the particular situation of WEEE management in the informal sector. A code of practice should cover, among other things, occupational health measures, best practices, formalization of the informal sector, and the formation of cooperatives.[51]

---

[49] S. Prakash, A. Manhart, Y. Amoyaw-Osei, and O. O. Agyekum, 'Socio-economic assessment and feasibility study on sustainable e-waste management in Ghana', Öko-Institut e.V. & Green Advocacy Ghana (2010) available at <www.oeko.de/oekodoc/1057/2010-105-en.pdf> (last accessed on 7 August 2015); A. Manhart, O. Osibanjo, A. Aderinto, and S. Prakash, 'Informal e-waste management in Lagos, Nigeria – socio-economic impacts and feasibility of international recycling co-operations', Institute for Applied Ecology and Basel Convention Coordinating Centre for Africa (BCCC-Nigeria) (2011) available at <www.basel.int/Portals/4/Basel%20Convention/docs/eWaste/E-waste_Africa_Project_Nigeria.pdf> (last accessed on 7 August 2015).

[50] S. Umair, 'Informal electronic waste recycling in Pakistan', Thesis, KTH Royal Institute of Technology (Sweden 2015) available at <http://kth.diva-portal.org/smash/get/diva2:813604/FULLTEXT01.pdf> (last accessed on 11 August 2015).

[51] Lundgren (n 48).

7.6   TRENDS AND OUTLOOK

Rapid innovation cycles and growing volumes of cheap EEE have brought about steep increases in the quantities of WEEE. Technological advances include the switchover to digital-only television in Europe, North America, and other industrialized regions of the world, which will accelerate the disposal of obsolete devices and stimulate trade in used EEE with developing nations. In addition, the material composition of EEE is tending to become more complex and the raw material supply more critical. Technologies to recover them from WEEE streams are needed, but these are increasingly complex and expensive. In addition, the past and current use of hazardous substances in EEE will shape WEEE management systems for a long time to come.

It is encouraging that legislation for sustainable WEEE management is rapidly being adopted in many countries. However, with the implementation and enforcement of new regulations still ahead, the main challenges are yet to be faced, especially in developing and transition countries. It will be key to ensuring a level playing field for all actors in order to make cannibalizing of WEEE solely for valuables impossible and to avoid harmful practices in WEEE recycling. In addition to existing waste policies and legal frameworks, WEEE-related regulations need to be enforced, likewise posing challenges to coordination between different regulatory bodies.

7.6.1   Increased Collection Rates and Improved Recycling Yields

Secondary resources are becoming more and more relevant given the shift of raw materials into products and the increasing demand for new raw materials. As outlined in this chapter, collection rates in industrialized countries are still far below their potential. Apart from illegal exports of EEE or WEEE to non-OECD countries, one reason for this is the lack of access to take-back schemes, which results in consumers storing EEE for longer periods of time and/or disposal of EEE through the municipal waste stream or scrap dealers. Higher collection rates need to be achieved in combination with improved recycling rates. In developing countries, most products enter the recycling chain through the informal sector, which is characterized by high collection rates. An international division of labour in WEEE recycling could link geographically distributed treatment options, combining pre-treatment at the local level and end-processing in state-of-the art facilities.[52]

---

[52]   As outlined in Wang et al (n 41).

In the future, the development of WEEE take-back schemes will also need to address technical and operational aspects of recovery of scarce[53] and critical metals.[54] The predominant technology in WEEE recycling is mostly oriented toward the recovery of base and precious metals, whereas scarce metals such as indium, gallium, germanium, and neodymium are lost in today's recycling system. The challenge to recycle a complex waste stream such as WEEE has to be addressed by appropriate recycling systems which are developed following a product-centric approach, especially for the case of metals: 'based on the holistic view of all elements contained in WEEE, it maintains and innovates a sophisticated physical and metallurgical processing infrastructure to produce high quality metals from complex multi-material recyclates'.[55]

This requires all stakeholders in the recycling chain (product designers, collectors and processors) to understand the whole system and the respective infrastructure to be adaptive to the changing composition of the waste. In analogy to the geological minerals processed by primary metallurgy, WEEE can be considered human-made minerals. Thus, the recyclers of complex modern products increasingly need the expertise of metal miners.[56]

In addition, a comprehensive international approach is required to ensure sustainable recovery of secondary resources. Among other elements, this might include harmonization of international standards toward fair recovery and trade of secondary resources and applying international financing mechanisms.

7.6.2 International Standards toward 'Fair' Secondary Raw Materials

Developing countries are suppliers of primary, but in recent years increasingly also of secondary raw materials. On the demand side, consumers in industrial countries are increasingly concerned about production circumstances of imported goods and wish to have transparent product declarations. While quality, social and environmental labelling is well established for some renewable commodities (e.g., Forest Stewardship Council labelling – (FSC)), it is almost non-existent for non-renewable

---

[53] B. J. Skinner, 'Earth resources' (1979) 76(9) *Proceedings of the National Academy of Sciences USA* 4212.

[54] L. Erdmann and T. Graedel, 'The criticality of non-fuel minerals: A review of major approaches and analyses' (2011) 45(18) *Environmental Science & Technology* 7620.

[55] UNEP, 'Metal Recycling' (n 32).

[56] Ibid.

commodities (one of the few examples is XertifiX – 'natural stone without child labour') and does not exist at all for non-renewable secondary commodities (e.g., precious metals from PWB recycling).

Environmental and social issues linked to informal and formal recycling also cause image problems for producers, usually multinational companies. As described in this chapter, many informal recycling processes involve low material recovery efficiency and risk contaminating commodities with hazardous substances. Hence efficient and sustainable recovery of secondary raw materials is a market opportunity that requires functioning 'reverse supply chains' with adequate capabilities for recycling and refining as well as sufficient monitoring of the quality of the recovered material in addition to scrutinizing the environmental and social impacts of the related processes. Therefore the harmonization of international standards and the introduction of processes to identify 'fair' secondary resources will be instrumental for leveraging these opportunities.

7.6.3 International Financing Mechanisms

Some of the substances potentially released by improper WEEE treatment are classified as POPs, ozone depleting substances (ODS), or greenhouse gases (GHG) and are regulated under international treaties such as the Stockholm Convention, the Montreal Protocol, and the Kyoto Protocol. Related to these are emission reduction schemes and/or international financing mechanisms, such as UN Environmental Finance Facility programs (e.g., Global Environment Facility – GEF), Cleaner Development Mechanisms (CDM), and voluntary systems (e.g., Verified Carbon Standard – VCS), which may be used for financing parts of processing WEEE properly to capture and destroy POPs and ODS. In addition, recovering secondary resources from WEEE as an alternative to mining primary resources can lower GHG emissions and is subject to the Cleaner Development Mechanism. Such international financing mechanisms might play a crucial role in implementing sustainable e-waste management systems by supporting initial investments as well as by creating market incentives to avoid improper processes and to remove internationally banned chemicals from the secondary resources market.

8. Landfill gas-to-energy as a contribution to greenhouse gas reduction

Jessica North[1]

EXECUTIVE SUMMARY

Landfills are found throughout the world and represent the prevalent method of waste disposal globally. Landfill gas, composed of approximately equal proportions of methane and carbon dioxide, is acknowledged as a significant contributor to greenhouse gas emissions. Methane is one of the short-lived climate pollutants requiring urgent action to mitigate. However, landfill gas also represents a potential source of 'green' power where it is extracted and combusted in a power generation facility. Landfill gas-to-energy projects therefore have the potential for a dual contribution to greenhouse gas reduction through mitigation of methane emissions and avoidance of fossil-fuel power. In addition, landfill gas extraction and combustion represents a key component of sustainable landfill management practices, essential for reducing the risk of gas migration and associated human and environmental impacts.

Given the available and proven technology, and the cross-benefits of improved landfill gas management, landfill gas-to-energy could be viewed as a 'low hanging fruit' for greenhouse gas mitigation. However, despite widespread adoption of landfill gas-to-energy projects in Northern European countries, North America, and metropolitan Australia and New Zealand, the majority of landfilled waste at the global level is not subject to gas capture and extraction systems. Barriers to growth in projects include technical limitations in some poorer regions, but are primarily due to weak regulatory environments and lack of financial incentives.

[1] Landfill Gas Industries Pty Ltd, Australia. The author sincerely thanks Dr Jean Bogner and Adam Bloomer for providing comments on early drafts of this chapter. The views expressed in this chapter remain those of the author alone and do not necessarily reflect the views of Landfill Gas Industries Pty Ltd, Australia.

Historically, the major drivers for development of landfill gas-to-energy projects have been regulatory requirements and revenue generated through a combination of power sales, carbon credits and/or renewable energy certificates. At both the international and national level, uncertainty in policies governing carbon and renewable energy markets, and the consequent market instability, have compromised the growth of investment in landfill gas to energy.

8.1 INTRODUCTION

In the majority of countries around the world, controlled and uncontrolled landfilling of untreated waste is the primary method of disposal. Methane (CH_4) emissions from landfill represent the largest source of greenhouse gas (GHG) emissions from the waste sector, contributing around 700 Mt CO_2-e to global emissions (estimate for 2009).[2] However, landfill methane also represents a potential fuel resource that can be harnessed to generate a form of 'green' energy where suitable conditions exist. The impact of landfill gas-to-energy (LFGE) projects on GHG reduction is therefore twofold: GHG is mitigated through the combustion of methane, and GHG-intensive fossil-fuel power is replaced by a 'green' power. Despite widespread installation of LFGE systems across northern Europe and North America, and many landfills in countries of the Organisation for Economic Co-operation and Development (OECD), there remains a considerable, untapped potential across the world. Technical, political, and financial challenges impede a more global uptake of landfill gas to energy projects.

This chapter is intended to provide a pragmatic overview of landfill gas (LFG) recovery and energy generation, including technical, environmental, regulatory and economic considerations. The author draws on experience of the Australian LFG industry as well as work with developing nations. The content is targeted at those who may be only marginally familiar with the LFG industry, and who wish to increase their understanding, for example as part of a project feasibility assessment or for policy development. In the current global context of reduced financial incentives for LFGE projects, this discussion may also assist partners in

---

[2] J. Bogner, M. Abdelrafie Ahmed, C. Diaz et al, 'Waste Management, in Climate Change 2007: Mitigation', Contribution of Working Group III to the Fourth Assessment Report of the Intergovernmental Panel on Climate Change, available at <www.ipcc.ch/pdf/assessment-report/ar4/wg3/ar4-wg3-chapter10.pdf> (last accessed on 22 March 2016).

developing nations to better understand the reasons why numerous Clean Development Mechanism LFG projects have been postponed or halted by project developers in recent years.

8.2   LANDFILL GAS

8.2.1   Generation

Waste contains organic material, such as food, paper, wood, and garden trimmings. Once waste is deposited in a landfill, microbes begin to consume the carbon in the organic material, which causes decomposition. Under the anaerobic conditions prevalent in landfills, the microbial communities contain methane-producing, or methanogenic, bacteria. As the methanogenic microbes gradually decompose organic matter over time, methane (approximately 40–60 per cent), carbon dioxide (approximately 30–50 per cent), and other trace amounts of gaseous compounds (< 1 per cent) are generated and form LFG. In controlled landfills, the process of burying waste and regularly covering deposits with a low permeability material creates an internal environment that favours methane-producing bacteria. As with any ecological system, optimum conditions of temperature, moisture, and nutrient source (i.e. organic waste) result in greater biochemical activity and hence greater generation of LFG.

The gradual decay of the carbon stock in a landfill generates emissions even after waste disposal has ceased. This is because the chemical and biochemical reactions take time to progress and only part of the carbon contained in waste is emitted in the year in which waste is disposed. Most is emitted gradually over a period of years. The actual rate of methane generation depends on numerous factors, including climate, cover and capping practices, leachate management, site size and depth, and site age.

Open, uncontrolled waste dumps (prevalent in developing regions) receiving moist, highly organic waste can still generate methane if there is a sufficient depth of waste mass to create anaerobic pockets (i.e. 5–10 metres). However, where landfill practices are informal and do not extend to site compaction and cover, the optimum anaerobic conditions for methane-production do not develop consistently across the site. Ironically, methane emissions increase as landfills become better managed, with impermeable liners, periodic use of cover material and compaction creating more anaerobic conditions within sites. Therefore, as developed countries make significant reductions of their landfill methane emissions, primarily through waste diversion and landfill gas capture, global emissions continue to increase as developing nations move away from open dumps and burning towards more controlled landfilling practices.

Methane is recognised as a GHG with a significant Global Warming Potential (GWP),[3] which has been re-evaluated by the Intergovernmental Panel on Climate Change (IPCC) from a value of 23 (in 2001)[4] to a proposed value of 34 (in 2013),[5] when a time horizon of 100 years is considered. As a relatively short-lived climate pollutant (i.e. around 12 years' duration in the atmosphere), the GWP of methane is much higher when a 20-year time horizon is considered (i.e. GWP of 86, according to the 2013 IPCC Fifth Assessment Report). This data points to the need for rapid implementation of methane abatement measures to address the immediacy of the climate change situation.

8.2.2   Emissions Estimates

Calculations for estimating emissions from decomposition of waste in landfill are subject to high levels of uncertainty. An accurate method for direct measurement of fugitive landfill emissions[6] is not currently available and therefore all estimates are based on theoretical models such as the IPCC First Order Decay model.[7] Available models are based on a number of underlying assumptions. Even if data obtained for waste quantities and composition are accurate, subsequent assumptions on decomposition rates, methane generation rates and oxidation rates among others, add error and uncertainty to the calculations. For example, there are numerous landfills known to be capturing 100 per cent or more of the methane estimated to be generated by the site using standard LFG models. These standard models also do not account for major drivers of methane

---

[3]   The ability of a substance to absorb infrared radiation and influence atmospheric warming is measured as its Global Warming Potential (GWP).

[4]   Intergovernmental Panel on Climate Change (IPCC), 'Third Assessment Report – Climate Change 2001', available at <www.grida.no/publications/other/ipcc_tar> (last accessed on 2 August 2015).

[5]   IPCC, 'Climate Change 2013: The Physical Science Basis', Contribution of Working Group I to the Fifth Assessment Report of the Intergovernmental Panel on Climate Change, available at <www.climatechange2013.org> (last accessed on 2 August 2015).

[6]   Fugitive landfill emissions are those emissions that exit the landfill, despite any LFG capture or oxidation system.

[7]   IPCC, 'Volume 5 Waste. Chapter 3, Solid Waste Disposal', in IPCC Guidelines for National Greenhouse Gas Inventories (2006) available at <www.ipcc-nggip.iges.or.jp/public/2006gl/pdf/5_Volume5/V5_3_Ch3_SWDS.pdf> (last accessed on 2 August 2015).

emissions, such as: a) the areal extent and physical properties of cover materials; b) the direct physical effect of LFG recovery, which lowers soil gas CH_4 at the base of the cover, reducing diffusive flux of CH_4; and c) seasonal CH_4 oxidation, which depends on seasonal climate (temperature and moisture) in site-specific cover soil profiles.[8]

The LFG industry tends to avoid reference to hypothetical LFG 'capture rates', which can be misleading and detrimental to sound LFG management. The rate at which a landfill will generate methane for extraction is most accurately gauged by expert investigation of the site, its history and management, and by drilling wells and examining gas flow data.

It should be noted that diversion of organic wastes from landfill and implementation of active systems for landfill gas extraction can be complimentary. In many OECD countries, diversion of paper, cardboard, food and/or garden materials from domestic waste has happily coexisted with successful landfill gas extraction systems, many generating power. Waste prevention and beneficial recovery of materials should be the long-term objective of all integrated municipal solid waste management systems.

8.3   LANDFILL GAS EXTRACTION

Landfill gas can be actively extracted from landfills and combusted to convert the methane to less harmful carbon dioxide. Passive systems also exist, where wells are installed to vent methane to the atmosphere, or through a 'biofilter', thereby reducing the risk of spontaneous landfill explosions and gas migration into neighbouring communities. Passive systems that include a 'biofilter' also achieve methane reduction through oxidation, and offer a further tool for control of landfill emissions. The present discussion addresses only active landfill gas extraction.[9]

An active extraction system may be installed in either closed or operating landfills. Vertical extraction wells are typically constructed by drilling down to near the base of the waste mass, inserting a perforated

---

[8]   K. Spokas, J. Bogner and J. Chanton, 'A Process-Based Inventory Model for Landfill CH_4 Emissions Inclusive of Soil Microclimate and Seasonal Methane Oxidation' (2011) 116 *Journal of Geophysical Research: Biogeosciences*, available at <http://onlinelibrary.wiley.com/wol1/doi/10.1029/2011JG001741/full> (last accessed on 2 August 2015).

[9]   A review of methane oxidation methods is provided in: C. Scheutz, P. Kjeldsen, J. Bogner, A. deVisscher, J. Gebert, H. Hilger, M. Huber-Humer and K. Spokas, 'Microbial Methane Oxidation Processes and Technologies for Mitigation of Landfill Gas Emissions' (2009) 27 *Waste Management and Research* 409.

high-density plastic pipe (i.e. polyethylene, 160 mm diameter), and surrounding the pipe with aggregate material to prevent waste materials fouling and blocking the pipe and allow gas to seep in to the pipe (see Figure 8.1 below).

Spacing of wells across the area of extraction may vary between 10 to 30 metres, depending in part on the primary objective of the system, such as odour and/or gas migration control, meeting a regulatory requirement, or maximising gas capture for power generation.

Horizontal, or lateral, wells can be progressively installed in active landfills, where waste is deposited in successive 'lifts'. Landfilled waste can begin producing methane gas within six months of deposition, if conditions are conducive to methanogenic activity, and combinations of lateral and vertical wells enable earlier extraction of gas. Lateral wells are fitted with horizontal sections of pipe, known as 'risers', which protrude from the waste mass and are eventually connected to the gas network. Protecting exposed risers from being damaged by landfill machinery is challenging.

There are effectively two styles of landfill gas system: the manifold approach, used primarily by UK and Australian operators, and the centralised well approach, used by Canadian and US operators. The manifold-type gas system consists of evenly spaced horizontal and lateral wells connected by lateral lines to a series of manifold stations, which are in turn connected to one or more combustion devices via a main gas line (see Figures 8.1 and 8.2 below).

Each well can be 'tuned' at the manifold stations to ensure an even flow of gas across the field. Landfill gas is moist, so lateral lines are installed on a gentle slope, with condensate traps to capture accumulated moisture and prevent blockages. Lateral lines and main lines are often trenched into the landfill, although lines may be left exposed on the surface of closed, capped sites to minimise excavation. Manifold-type systems may be more applicable for landfill sites in developing nations, due to the lower cost of installation (equipment and materials) and the ability to more finely tune gas flows across a potentially highly variable gas field.

The centralised well approach to gas system design, common in the US, is to install larger diameter vertical wells at wider intervals, with no lateral lines and no manifold stations. Flows are then regulated at each well head, and each well head is directly connected to the main gas line. This approach may be particularly effective for larger, very productive gas fields.

Gas quality must be monitored regularly: methane (CH_4), carbon dioxide (CO_2), and oxygen (O_2) concentrations provide an indication of the stability of the gas field. A gas field can be over-extracted, which

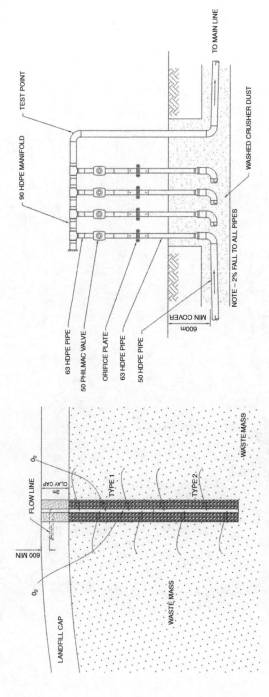

Source: Landfill Gas Industries Pty Ltd, Australia.

Figure 8.1 Schematic of vertical, top extraction well and gas field manifold station in a manifold-type landfill gas system

177

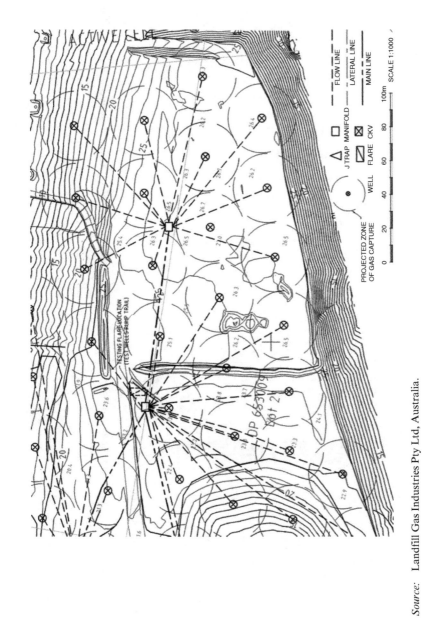

Source: Landfill Gas Industries Pty Ltd, Australia.

Figure 8.2   Example of a landfill gas field design

impairs (or even stops) methanogenic activity, and excessive ingress of O_2 can lead to explosive ratios of methane and subsequent landfill fires.

Landfill gas extraction systems can be retrofitted to existing, active, unlined landfill sites – this is not uncommon practice in many developed countries, where currently active landfills may have started operation before regulations required liners or LFG management. Modifications to well and gas field design can reduce oxygen ingress and improve capture efficiencies, particularly for sites which have had limited compaction and/ or cover material emplaced. There is no 'one size fits all' design for landfill gas systems, and each site requires careful assessment by an experienced LFG operator to determine the best design to meet the site's needs.

8.4 LANDFILL GAS COMBUSTION

Once LFG is able to be extracted through the network of pipes, the system must be fitted with a device capable of combusting the gas and thereby converting methane to less harmful carbon dioxide. Combustion may be achieved by flaring and/or a gas engine with consequent generation of power. Combustion devices have been specifically designed to cope with the 'dirty' (i.e. impure and corrosive) nature of landfill gas. Methane in landfill gas provides the hydrocarbon fuel for combustion in devices. It should be noted that there are examples of LFG projects where heat has been produced, as opposed to electricity. The present chapter focuses on the more common objective of electricity generation.

8.4.1 Flaring

Flares require a source of external power to operate their 'blowers' – LFG fields are not under high pressure, and require a gentle vacuum to extract the gas from the site. Flares are available in a range of sizes, and can be used to combust as little as 50 cubic metres of landfill gas per hour (50 m^3 LFG/h) up to 1,000 or more m^3 LFG/h. Multiple combustion devices can be connected to a single gas field, and a flare is usually installed as an emergency, back-up measure where a gas engine is being operated.

There is comparatively little variation in LFG flare technology – form largely follows function, and country-specific standards govern the manufacture of flare units to ensure safe operation and efficient combustion. Flares are typically classified as either 'open' ('candlestick') or 'enclosed' flares. Open flares burn LFG as an open flame – combustion occurs at the flare tip, which is elevated above ground. Since there is limited ability to control combustion, the necessary high combustion temperatures cannot be achieved to ensure consistent, high levels of destruction efficiency for methane and other volatile organic compounds. Open flares are becoming less common as regulations governing land-fills around the world become more stringent. In an enclosed flare, gas combustion occurs at the burner tip due to the secondary and controlled mixture of air to the burner mix. As the flame propagates upwards it consumes air from the natural draft to complete combustion. Combustion is completed in a controlled environment allowing for proven destruction efficiency and more reliable operation.

Landfill gas is typically captured and flared for several months prior to installing an LFGE facility on the site. Gas field operators will observe the trends in gas generation during this trial period and determine whether there is a sufficiently constant flow and quality of gas to operate a viable LFGE plant.

8.4.2 Landfill Gas-to-Energy

Most developed nations that have continued to rely on landfilling for waste disposal, such as the US, Canada, Australia and New Zealand, have developed LFGE. EU Member States with large quantities of 'legacy' landfills (e.g. UK, southern and eastern Europe) as well as emerging economies and developing countries (e.g. South Africa, Mexico, South and Central America, and selected Asian countries) also have experience with LFGE, often through joint ventures with developed countries.

The decision to opt for power generation is complex, and depends on factors such as the bureaucracy governing connections to power grids or proximal off-take partners for power, available funding (power facili-ties require comparatively large capital expenditure), available technical expertise (power facilities require on-call technicians), financial incentives for generating green energy (i.e. mandatory renewable energy targets for power retailers, renewable energy certificates, and feed-in-tariffs), and stability and longevity of the project. In developed countries, 20–30 year contracts granting landfill gas rights to power facility operators are not uncommon – the payback period for a multi-million US dollar investment in a gas field and power generation infrastructure can be lengthy.

8.4.3 Landfill Gas-to-Energy Technology

Power can be generated from LFG through various technologies, including the reciprocating internal combustion engine (RICE), gas turbine or steam turbine engines, direct use of LFG in a gas-fuelled boiler (creating steam or hot water for industrial uses), micro-turbines on small LFGE projects, and upgrading LFG to a natural gas quality. Table 8.1 provides a high-level comparison of the main types of LFG to energy (LFGE) technology options. Globally, the majority of LFGE projects involve the generation of electricity, hence the focus of the following discussion.

In general, RICEs have proved to be the most cost-effective and reliable technology for electricity generation from LFG, especially for moderately sized projects, generating less than 5 MW. However, although RICEs have a comparatively low capital cost per kilowatt (kW), they have higher operation and maintenance costs than gas turbines.

LFG flaring units and power generation plants have some technical components in common, such as the condensate knockout vessel, valve controls, blower, flame arrester and automatic block valve, air intake louvres, monitoring ports, and control panel. However, when gas exits the compressor stage, it generally requires additional treatment prior to entering the engine. Treatment includes additional particulate filters, and chiller systems to reduce the temperature of the gas and bring it to dew point (formation of water droplets). By inducing droplets of moisture to form in this manner, approximately 90 per cent of water-soluble contaminants are removed. A secondary knockout vessel, and a Siloxane removal system, can also be considered. Siloxane removal is an expensive treatment process and is generally considered too expensive to install and maintain. However, as the sources of Siloxane increase in landfilled waste (such as cosmetics, hair products, deodorants, and lubricants), Siloxane treatment may be increasingly necessary to enable LFGE projects.

The gas engine and ancillary equipment are usually housed in a container unit that is acoustically designed for noise reduction and insulated to maintain suitable ambient temperature. Exhaust air may pass through a silencer to reduce noise emissions prior to release to the atmosphere. In some cases, secondary post-combustion exhaust gas treatment is required where emission limit values are more difficult to achieve for site specific gas qualities.

A back-up flare is also required adjacent to the power facility, to enable gas destruction during periods of scheduled and un-scheduled shut down. Telemetry systems are also becoming standard practice on power facilities, enabling operators to monitor, diagnose, and trouble-shoot remotely.

Engines range in size, output, efficiency, and can have varying sensitivity to levels of methane – higher methane concentrations in the incoming landfill gas result in better power output from the engine. Lower methane concentrations mean that a higher flow rate of landfill gas will be required to achieve the same power output. Some gas engine models require as little as 200 m^3 LFG/h (at 50 per cent methane content) to generate 400 kW of

Table 8.1 Comparison of several LFG-to-power technologies

	RICE	Gas Turbine Engine	Steam Turbine Engine	Gas-fuelled Boiler (Steam/Hot Water)	Micro-turbine	LFG Upgrading
Advantages	Lowest cost option per MW installed. Relatively robust, available in incremental sizes.	Lower running costs (in theory). Good variable output ability.	Simple design. Isolation of fuel from motivator, hence low maintenance.	Lowest cost option for beneficial use of LFG.	Can be considered for sites with very low gas flow (i.e. 10–100 kW output).	A potentially viable option where natural gas prices are high, and LFG qualifies for renewable energy credits.
Disadvantages	Preventative maintenance is key to good availability on RICE technology. Average air emissions. Oil consumption can be high.	High capex, specialised technology. Not as efficient as RICE technology. Does not cope with Siloxane in landfill gas. Units de-rate in higher ambient temperatures (i.e. tropical climates).	More components, higher cost per MW capex than RICE technology, the least efficient of all three due to the heat exchanging process. Steam handling requires specialised skill.	Requires an off-take partner for the steam or hot water produced, and in close proximity to the landfill.	Gas may require pre-processing to achieve sufficient quality. Project may only be commercially viable where a direct off-take partner is located in close proximity to the facility.	Requires costly and energy-intensive gas treatment to achieve quality of natural gas, and compression to input to natural gas main lines or bottle.

power. In developed countries, gas engines are typically 1MW or larger units, requiring at least 550 m^3 LFG/h (at 50 per cent methane content).

8.4.4   Viability of Landfill Gas-to-Energy

Several key factors should be evaluated when considering the overall (practical) viability of a LFGE project, specifically for electricity generation. These include electrical conversion efficiency, reliability (of equipment and gas field), system flexibility, and electricity supply infrastructure.

Electrical conversion efficiency is an indication of what portion of the energy value of the landfill gas can be converted into electrical power. Electrical conversion efficiency varies based on the selected technology. For example, internal combustion engines have a higher efficiency than most gas turbines; however very high altitudes or high ambient temperatures reduce their efficiency.

Reliability of the equipment and the supply of the fuel to the LFGE plant will determine the actual amount of power generation. The need for spare gas engine parts must be assessed based on the availability of these parts in the specific country, as well as the time that may be required to import the parts. Operating the LFGE plant in accordance with equipment specifications and conducting regularly scheduled maintenance will reduce the wear on system parts and allow plant operators to plan for outages, thereby reducing plant downtime.

The modular nature of internal combustion engines and gas turbines provides flexibility for incremental capacity increases in response to greater production of landfill gas. Internal combustion engines or micro turbines can be added in smaller incremental stages than gas turbines for a lower capital cost.

Power generated by a LFGE project is often transmitted to a local power grid and sold as a form of revenue. Power is exported via a step transformer to the local distribution network. A grid connection and load study needs to be completed to anticipate potential voltage rise and the requirements of the unit to import and export volt-amperes reactive (VARs). Typically, LFGE projects rely on existing infrastructure to deliver electricity to the market because the costs of building extensive new infrastructure are prohibitive. The project developer must examine the availability and types of nearby power lines and electrical substations. Nearby power lines that are suitable to provide a connection to the power grid and substations are advantageous for project development. Interconnection can be a considerable investment cost and requires careful investigation into permits and approvals that can vary greatly, depending on the location and site-specific requirements.

LFGE project economics are highly site-specific, and often involve multiple partners (e.g. landfill owners, operators, LFG developers, LFG customers, utility companies, local government, etc.).

8.4.5 Landfill Gas-to-Energy and GHG Reduction

LFGE projects mitigate GHG emissions by converting methane to CO_2 through combustion, as well as avoiding the GHG emissions associated with power derived from fossil fuels. The 'net abatement' achievable by LFGE takes into account any power imported to the facility, for example during shut-down and maintenance periods, and any ancillary fossil fuel used. Net abatement will vary depending on a range of factors, particularly technology. As an example, a 1 MW rated RICE, consuming 550 m³/h of LFG (50 per cent methane), with 95 per cent availability, could directly mitigate approximately 2,288,550 m³ (i.e. 1,553 t) of methane per year. The calculated carbon intensity of electricity derived from the standard grid varies between and within countries – in Australia, for example, estimated values range from 0.20–1.17 kg CO_{2-e}/kWh, with higher values reflecting a greater proportion of grid electricity produced by coal-fired power stations. Therefore, a 1 MW LFGE facility operating in Australia could be estimated to avoid between 1,664 and 9,737 t CO_{2-e} each year, depending on location. When the dual impact of LFGE on GHG emissions is viewed at a global scale, the magnitude of potential, achievable GHG reduction is considerable.

8.5 KEY DRIVERS FOR LANDFILL GAS-TO-ENERGY

LFG extraction and combustion systems are relatively costly to install, operate and maintain, and landfill owners must therefore be compelled to procure a system through regulatory requirement and/or financial benefit. LFGE project costs include capital and labour costs to purchase and install the equipment needed to treat the gas and generate electricity, as well as on-going operation and maintenance costs. These latter costs should not be underestimated and include labour and materials used to operate the system and perform routine maintenance and repairs, such as periodic equipment overhauls. In addition, proximity to existing power infrastructure can be critical – project development costs may escalate if power poles and lines must be extended a considerable distance to reach an LFGE facility. For example, in Australia, the feasibility studies and upgrade works required for an interconnection can cost USD 400,000 or more (and require more than 12 months to achieve). As an indicative figure, in the Australian context, the installation of a gas field, flare and a 1 MW output gas engine could cost in the range of USD 2.5–4 million, with annual operation and maintenance costs of 5–10 per cent of establishment cost.

8.5.1 Regulation

The structure of Australian landfill gas policy and regulation has comparable elements to the policy and regulatory structures in the EU and US, and is provided in the present chapter to illustrate widely applicable possibilities and challenges for LFGE.

Policies directing landfill management in Australia are applied at the State level, rather than at the Federal, or Commonwealth, level of government. Each State or Territory is responsible for preparing guidelines for proper operation of landfills, including LFG management. The guidelines enable State-based Environmental Protection Agencies (EPAs) to set licence conditions for landfill sites and determine whether sites comply with these conditions. Licences vary on a site-by-site basis, with considerable variation in the interpretation of guidelines applied to sites – States such as Victoria have introduced increasingly prescriptive guidelines to standardise landfill practices (i.e. EPA Victoria's *Best Practice Environmental Management* publication for siting, design, operation and rehabilitation of landfills (*Landfill BPEM*)). LFG, as a key component of landfill management, is referred to in guidelines, with varying degrees of detail. Similarly, the EU Landfill Directive provides a limited set of general requirements for control of LFG[10] and Member States implement individual regulations, which in some cases are far more stringent and prescriptive (e.g. the regulations of UK, Norway and Sweden).

The requirement for active LFG extraction and combustion tends to be specified in landfill licences for sites operating in urban areas, whereas regional Australian landfills can often have no regulatory (licence) requirement specifically referring to gas extraction. Regional Australian landfills are predominantly owned and operated by local government, with a large range of sizes and levels of management sophistication. Most regional sites accept in the range of 10,000 to 100,000 tonnes of domestic and commercial waste each year. Despite evidence of gas generation in the majority of landfills accepting putrescible waste, local governments are unlikely to allocate resources for active gas extraction unless required to do so by

---

[10] European Union Council Directive 1999/31/EC, Annex 1, Paragraph 4.

prescriptive licence conditions. Due to a number of factors, such as the composition of waste and climate, landfills receiving as little as 40,000 tonnes of domestic waste per year can be technically viable sites for LFGE projects in Australia. The combination of tighter regulatory control on LFG management and incentives for investment in 'green' energy could see a significant increase in LFGE projects developed in Australia, and globally, with associated mitigation of greenhouse gas.

8.5.2 Power Demand/Prices and Incentives

LFGE projects will not occur without sufficient financial incentive, which can potentially be provided by three revenue sources: wholesale power sales; certificates for renewable energy; and credits for carbon offsets. Wholesale power prices are driven by high demand and/or low supply. Australia's energy market is dominated by comparatively cheap coal-fuelled power, and a lower than expected demand for power in recent years. Australian LFGE projects are therefore reliant on additional income from renewable energy certificates, which are saleable under the Government's Renewable Energy Target. The ability for projects to create and market renewable energy certificates and carbon credits depends on government policy and regulatory mechanisms. LFGE represents a particularly desirable source of green energy, given that it provides base-load power (i.e. power is generated 24 hours a day, during both peak and off-peak usage periods), unlike solar and wind power, which fluctuate with the availability of sun and wind. However, in recent years Australia's renewable energy and carbon policies have been in a state of flux, creating uncertainty in markets for both renewable energy certificates and carbon credits.

Historically, Australian LFGE proponents have negotiated long-term (i.e. ten or more years) power purchase agreements (PPAs) with energy retailers, often combining power and renewable energy certificates, which have provided commercial certainty to invest in projects. Due to fluctuating power demand and prices, and an uncertain future for Australia's Renewable Energy policy, energy retailers are not offering comparable long-term PPAs. The result is a delayed uptake of LFGE projects across Australia, and a loss of a viable and reliable source of green energy, as well as GHG mitigation. The Australian experience is not dissimilar to the global situation, post-Kyoto.

8.6 GLOBAL CHALLENGES FOR LANDFILL GAS-TO-ENERGY

Apart from particularly proactive countries such as many northern EU Member States, the lack of regulatory drivers for active LFG capture and combustion is widespread. The necessary financial incentives, such as green energy certificates or carbon credits, are also largely absent around the world. Practical, technical obstacles further frustrate a more global adoption of LFGE.

For several years (i.e. 2004–12), the Clean Development Mechanism (CDM) provided the necessary stimulus for instigation of LFG and LFGE projects in developing nations. The CDM was established under the Kyoto Protocol to allow a country with an emission-reduction or emission-limitation commitment under the Kyoto Protocol (Annex B Party) to implement an emission-reduction project in a developing country. In the international negotiations to determine a post-Kyoto agreement, there appears to be widespread support for continuing the CDM. Approved projects can earn certified emission reduction (CER) credits, each equivalent to one tonne of CO_2, which could be sold and counted towards meeting international emissions reductions targets.[11] In past years, the CDM removed the financial barrier for some developing nations to implement LFG capture and combustion projects, including LFGE. Landfill gas projects represented a large portion of registered CDM projects (approximately 11 per cent), and recovered a reported 30 Mt CO_2-e of landfill methane in 2008.[12] China, Brazil and a number of other South American countries have dominated the list of LFG projects registered under the CDM (Figure 8.3).

However, the mechanism is now challenged by low prices for CERs, which have collapsed by more than 95 per cent since peak prices were reached in 2008. Until national commitments to reduce GHG are strengthened under a post-Kyoto agreement, or series of agreements, and global demand for CERs increases, the CDM no longer provides sufficient financial incentive to stimulate (or, in some cases continue) LFG projects in developing countries. Since the end of the Kyoto Protocol's first commitment period in 2012, very few LFG projects have registered under the CDM (see Figure 8.4).

---

[11] See <http://cdm.unfccc.int>.
[12] S. Monni, R. Pipatti, A. Lehtilla, I. Savolainen and S. Syri, 'Global climate change mitigation scenarios for solid waste management', Technical Research Centre of Finland (VTT Publications, Espoo 2006).

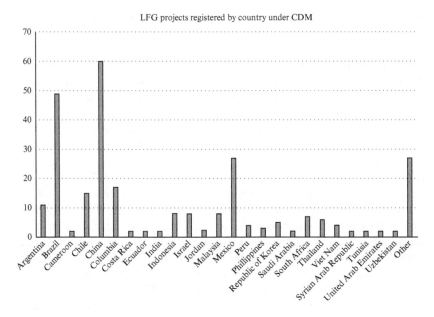

Source: 2014 data, CDM UNFCC Project Cycle database.

Figure 8.3   LFG CDM projects registered by country

There are currently 279 registered LFG projects under the CDM, but only 43 per cent of these have been issued CERs, and of the remaining 57 per cent of projects, the majority have not requested or been issued CERs since 2012 (see Figure 8.5). A number of these projects are believed to be currently on hold due to the low value of CERs. For example, several proposed LFGE projects in Malaysia, Vietnam and Bangladesh have not been developed since registering under the CDM. Evidently, a market-based tool like the CDM has the ability to significantly and rapidly incentivise LFGE projects around the world, but it relies on long-term market stability and certainty to maintain momentum. Once again, a parallel can be drawn between the global experience with the CDM and Australia's current, uncertain domestic carbon market, as previously presented. The expansion of LFGE projects largely depends on private sector investment, and the private sector requires certainty to invest.

Technical barriers include the complexity of connecting LFGE systems to the local energy grid, especially in regions where the electricity supply is highly controlled or monopolised. Unstable grids prone to frequent power outages will impact the efficient operation of LFGE facilities – engines will shut down during a power outage and require staff to inspect the unit

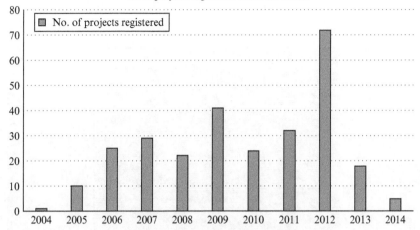

Source: 2014 data, CDM UNFCC Project Cycle database.

Figure 8.4   LFG projects registered under the CDM

prior to restarting. Although a considerable amount of system monitoring can be achieved remotely via telemetry systems, skilled operators are still required to manage gas fields and power facilities. The availability of suitably trained personnel can also determine the viability of LFGE. In developing countries, landfill operators may require education and technical assistance to improve landfill management and engineering even prior to consideration of a LFGE project. In India, for example, a major barrier to LFG management is a lack of awareness of the potential harmful impacts of unmanaged gas, as well as lack of relevant technical expertise in municipal waste departments.[13]

8.7   THE FUTURE OF LANDFILL GAS-TO-ENERGY

LFG capture and combustion is in many aspects a global 'low hanging fruit' for GHG mitigation: the necessary technology is available and proven, the cross-benefits for human and environmental protection are

---

[13]   Siddiqui et al, 'Review of Past Research and Proposed Action Plan for Landfill Gas-to-Energy Applications in India' (2013) 31(1) *Waste Management & Research* 3.

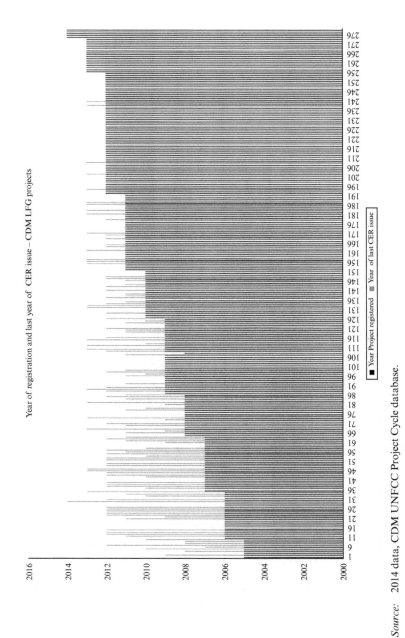

Year of registration and last year of CER issue – CDM LFG projects

■ Year Project registered   ▨ Year of last CER issue

Source: 2014 data, CDM UNFCCC Project Cycle database.

Figure 8.5   Year of registration of LFG projects under the CDM and most recent year of CER Issuance

demonstrable, and there is a plethora of landfills worldwide that do not currently extract and combust gas. Furthermore, the short-lived, potent nature of methane as a GHG should act as a major global stimulus to install LFG systems on these untapped landfills as soon as possible. Unfortunately, a combination of weak regulations and financial disincentives prevails, and the 'fruit' remains largely unpicked.

Experience in developed countries indicates that, increasingly, gas capture and combustion will be regulated and mandatory on landfills receiving putrescible waste in excess of threshold amounts. In reality, this increased regulatory environment may take many years to extend to developing countries, where a multitude of critical health and sanitation issues take priority, and resources to implement regulations are severely limited. Ultimately, government regulations need to be strengthened around the world to require the appropriate management of gas[14] from landfills receiving putrescible waste. Despite the impact that such regulations may have on the eligibility of LFG projects under carbon credit schemes such as the CDM,[15] the potential deleterious impacts of LFG on human health and the environment should dictate mandatory control.

The fate of a number of LFGE projects in developing nations would appear to depend on the future value of CERs, which in turn depends on a tightening of global, binding emissions targets. International, as well as domestic, carbon credit programmes have been shown to stimulate investment in LFGE; however, long-term stability in carbon markets is needed if that investment is to continue. Unfortunately, where regulatory requirements for LFG control do not exist (or are not enacted) and financial incentives are significantly diminished or removed, LFG projects will be discontinued.

With increasing global interest in renewable energy, it is likely that national renewable energy targets and incentives will continue to develop and strengthen around the world. As a base-load source of green energy, LFGE has the potential to contribute to the stability of a diversified, renewable-based power supply. Carbon policy worldwide needs to recognise the dual impact of LFGE on GHG reduction, which should

---

[14]   Appropriate management of LFG may include active extraction and combustion, oxidation systems, and/or passive ventilation methods, depending on site size and situation.

[15]   See the discussion in UNEP, 'Waste and Climate Change, Global Trends and Strategy Framework' (UNEP 2010) available at <www.unep.org/ietc/Portals/136/ Publications/Waste%20Management/Waste&ClimateChange.pdf> (last accessed on 2 August 2015).

enable projects to claim carbon credits for methane abatement, and renewable energy certificates for avoidance of GHG emissions from fossil-fuel-based power sources. As with any industry, the LFG sector needs a level of long-term commercial and regulatory certainty in order to invest in projects and continue to make a major contribution to GHG reduction.

9. Opportunities for economically and environmentally sound energy and resource recovery in Asia

Jinhui Li, Xiaofei Sun and Baoli Zhu[1]

EXECUTIVE SUMMARY

Although the increase of solid waste generation is a considerable problem faced by the whole world, it is more severe in Asian countries owing to their rapid urbanization and industrialization over the past few decades, especially in large developing countries such as China and India. In order to solve the problems relating to the energy shortage and the rapid growth of resource consumption, and to address the environmental pollution caused by solid waste generation, more and more countries in Asia are focusing on energy and resource recovery from waste. This chapter introduces the current status of waste generation and recycling in selected Asian countries, and discusses the existing problems and challenges in waste management and recycling. It has been found that increasing population and economic development not only contribute to the sharp rise in solid waste generation but also to its increasing complexity and hazardousness. In contrast to the selected developed and developing countries in Asia, the overall development of waste recycling is not balanced. Because of the backward technologies, environmentally sound solid waste disposal levels and resource recovery rates of solid waste in Asia is very low. Nowadays, the awareness of the public and of governments of solid waste management and recycling is rising; policies and regulation systems related to solid waste have been established; and new technologies (e.g. waste incineration power generation, biomass fuel, etc.) are being developed. The chapter concludes that energy and resource recovery in Asia has tremendous market potential in the future decades.

---

[1] The views expressed herein are those of the authors and do not necessarily reflect the views of the United Nations.

9.1   INTRODUCTION

Asia is the Earth's largest and most populous continent, located primarily in the Eastern and Northern hemispheres. It covers 8.7 per cent of the Earth's total surface and comprises 30 per cent of its land area. With approximately 4.3 billion people, it hosts 60 per cent of the world's current human population. In the past few decades, Asia has had a high growth rate in population and economics. In the twentieth century, Asia's population nearly quadrupled. Modern Asia has the second largest nominal GDP of all continents, after Europe, and the strongest purchasing power in the world. Rapid population and economic growth have led to severe problems related to the expanded consumption and depletion of resources, and the increased generation of wide-ranging types of waste in Asia, especially in large developing countries such as China and India.[2]

The large quantities of solid waste not only cause serious pollution to the environment, but also restrict the sustainable development of the economy in most Asian developing counties. When not properly treated, waste will have great impacts on human health and the environment (soil, water and air).[3] Many studies indicate that people living near waste disposal sites are negatively affected.[4] According to a report of UNEP,[5] the decay of organic waste contributes 5 per cent of greenhouse gases globally. Many materials containing rare resources are discarded as waste, for example e-waste, which is a huge economic and resource cost for the whole society.

But waste is not only a challenge, it is also a great opportunity. Proper

---

[2]   Amit Ray, 'Waste management in developing Asia: Can trade and cooperation help?' (2008) 17(1) Journal of Environment & Development 3.

[3]   Syeda Maria Alia, Aroma Pervaiza, Beenish Afzala, Naima Hamida and Azra Yasminb, 'Open dumping of municipal solid waste and its hazardous impacts on soil and vegetation diversity at waste dumping sites of Islamabad city' (2014) 26(1) Journal of King Saud University – Science 59; Nanna I. Thomsen, Nemanja Milosevic and Poul L. Bjerg, 'Application of a contaminant mass balance method at an old landfill to assess the impact on water resources' (2012) 32(12) Waste Management 2406.

[4]   Hongmei Wanga, Mei Hana, Suwen Yanga, et al, 'Urinary heavy metal levels and relevant factors among people exposed to e-waste dismantling' (2011) 37(1) Environment International 80; Elena De Felipa, Annalisa Abballea, Francesco Casalinoc et al, 'Domenico serum levels of PCDDs, PCDFs and PCBs in non-occupationally exposed population groups living near two incineration plants in Tuscany, Italy' (2008) 72(1) Chemosphere 25.

[5]   UNEP, 'Guidelines for National Waste Management Strategies: Moving from Challenges to Opportunities' (UNEP 2013) available at <www.unep.org/ietc/Portals/136/Publications/Waste%20Management/UNEP%20NWMS%20English.pdf> (last accessed on 7 August 2015).

waste management, recycling and reuse of waste will avoid the negative impacts associated with waste generation, provide an opportunity to recover resources, realize environmental, economic and social benefits and contribute to safeguard sustainable development in economy and society.

9.2 STATUS OF WASTE GENERATION AND RECYCLING IN ASIA

9.2.1 China

Unlike in the developed countries, solid waste management in China started relatively late. Along with the population growth and faster industrialization and urbanization, solid waste generation in China has continued to increase. The volumes of solid waste generated annually increased from 1020 million tons in 2001 to 3390 million tons in 2012, and the average annual growth rate was more than 11 per cent.[6]

Industrial solid waste is the most important stream of solid waste in China. With a growth rate higher than that of municipal solid waste (MSW) and hazardous waste, it accounted for 94 per cent of all solid waste in 2012, an increase from 87 per cent in 2001.[7] Figure 9.1 shows the volumes of industrial solid waste generation and disposal in China from 2003 to 2012. The volumes of harmless disposal and comprehensive utilization have increased along with the waste generation, although in 2011 and 2012, the utilization rate and disposal rate showed a relative decline (mainly due to the fast increase of industrial waste in 2011 and 2012).

The total amount of bulk industrial solid waste generated in China in 2010 reached 2.8 billion tons. At the end of the '11th Five-Year Plan'[8] period in China, the bulk industrial solid waste comprehensive utilization reached 1.1 billion tons (increased by 5600 million tons compared to the '10th Five-Year Plan' period);[9] the comprehensive utilization rate reached

---

[6] Analysis of solid waste pollution prevention and control industry development in China in 2014 (in Chinese) available at <www.askci.com/news/201406/16/1615385240474.shtml> (last accessed on 7 August 2015).

[7] Ibid.

[8] National People's Congress of the People's Republic of China, '11th Five-Year Plan for national economic and social development in China (2006–2010)' (in Chinese), available at <www.npc.gov.cn/wxzl/gongbao/2001–03/19/content_5134505.htm> (last accessed on 31 August 2015).

[9] National People's Congress of the People's Republic of China, '10th Five-Year Plan for national economic and social development in China

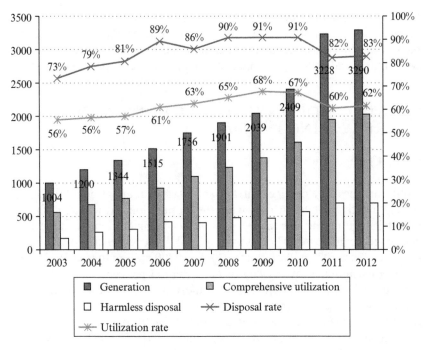

Source: Chinese statistical yearbook (2004–13).

Figure 9.1   Industrial solid waste generation and disposal in China, 2004–
13 (million tons)

40 per cent.[10] More than 15000 enterprises engaged in industrial solid
waste utilization, with an output value of 300 billion Yuan, and giving
employment to nearly 2 million.[11]

With the development of the national economy and the increase of the
population, the amount of MSW in China maintained a high growth rate.
In 2004, Chinese cities generated about 190 million tons of solid waste,
and China has earned the distinction of being the world's largest MSW

(2001–2005)' (in Chinese), available at <www.npc.gov.cn/wxzl/gongbao/2006-
03/18/content_5347869.htm> (last accessed on 31 August 2015).
   10   Ministry of Industry and Information Technology of PRC, '11th Five-Year
Plan on the comprehensive utilization of bulk of industrial solid waste' (2012)
(in Chinese) available at <www.miit.gov.cn/n11293472/n11293832/n11293907/
n11368223/n14484357.files/n14484192.pdf> (last accessed on 11 August 2015).
   11   Ibid.

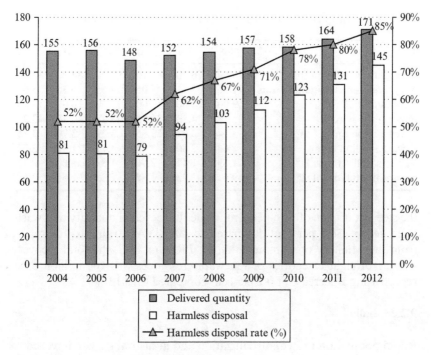

Source: Chinese statistical yearbook (2005–13).

Figure 9.2 MSW collection and disposal in China, 2005–13 (million tons)

generator, ahead of even the US.[12] It is estimated that by 2030, the amount of MSW will increase to about 480 million tons.[13] The amount of MSW collection and disposal in China from 2004 to 2012 is shown in Figure 9.2. In 2012, the delivered quantity of MSW in China reached 171 million tons from 155 million tons in 2004, and the harmless disposal rate of MSW had increased significantly in the last nine years.

Industrial renewable resource recycling in China started in 1997, and is still at an initial stage. Table 9.1 shows the main types of renewable resource recycling by volume in China from 2008 to 2013.[14] In 2013, plastics suffered

---

[12] Analysis of solid waste pollution prevention (n 6).

[13] Ray (n 2).

[14] Ministry of Commerce of PRC, National Development and Reform Commission, Ministry of Land and Resource of PRC et al, 'Renewable resources system construction medium and long-term plan (2015–2020)' (2015) 33(1) China Resources Comprehensive Utilization 3 (in Chinese).

Table 9.1   Main type of renewable resources recycling volume in China

Type	2008	2009	2010	2011	2012	2013
Iron and steel/million tons	70.60	76.20	83.10	91.00	84.00	85.70
Non-ferrous metals/million tons	1.96	3.61	4.05	4.55	5.30	5.62
Plastics/million tons	9.00	10.00	12.00	13.50	16.00	13.66
Papers/million tons	31.28	34.23	36.95	43.47	44.72	43.77
Tyres/million tons	3.14	3.07	3.35	3.29	3.70	3.75
E-waste/million tons	2.60	2.80	2.84	3.71	1.91	2.64
Vehicles/million tons	1.65	1.47	2.76	1.83	2.00	2.77
Steamship/million LDT	69.4	3.23	1.87	2.25	2.55	2.50

the steepest decline of 14.6 per cent, while the recycling quantity of e-waste increased by 38.3 per cent compared to 2012.[15] According to the report from the Ministry of Commerce of the PRC, in 2013, the total recovery volume of eight categories of renewable resources was 160 million tons, representing a decrease of 0.2 per cent compared to 2012.[16]

9.2.2   India

Rapid population growth, urbanization and industrial growth have led to severe waste management problems in the cities of developing countries such as India. Solid waste generated in India consists of municipal solid waste, plastics waste, construction and demolition waste, packaging waste, biomedical waste, e-waste and hazardous waste.[17] Nowadays, the majority of cities in India are not able to dispose of the enormous quantity of waste, and about 90 per cent of waste is disposed of by open dumping.[18]

There is no specific statistical data on solid waste generation in India. The Ministry of Urban Development of India assessed MSW generation in the country to be 0.1 million metric tons per day in the years 2001–02.[19]

---

[15]   Ministry of Commerce of PRC, 'Renewable resources industry development report in China' (2014) (in Chinese) available at <http://images.mofcom.gov.cn/ltfzs/201406/20140618113317258.pdf> (last accessed on 11 August 2015).

[16]   Ibid.

[17]   Ministry of Environment and Forests, 'Report of the Committee to Evolve Road Map on Management of Waste in India' (2010) available at <www.moef.nic.in/sites/default/files/Roadmap-Mgmt-Waste.pdf> (last accessed on 11 August 2015).

[18]   Tapan Narayana, 'Municipal solid waste management in India: From waste disposal to recovery of resources?' (2009) 29 Waste Management 1163.

[19]   Ministry of Environment and Forests, 'Report' (n 17).

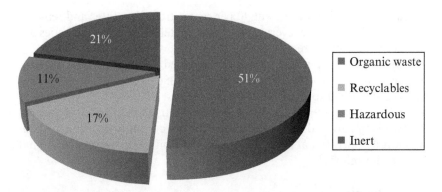

Figure 9.3   Typical composition of municipal solid wastes in India

And it is estimated that the MSW generation in India was 0.573 million metric tons per day in the year 2008.[20] The Federation of Indian Chambers of Commerce and Industry (FICCI) conducted a survey among Municipal Corporations of 48 cities to assess the management of solid waste in Indian cities. The survey showed that the highest waste generation per day was in the city of Delhi (6800 tons per day) and the lowest in Shimla (65 tons per day).[21] The typical composition of municipal solid wastes in Indian is shown in Figure 9.3. As can be seen, the proportion of organic waste in MSW in India is very high, but there are only 110 facilities in the country which manage to treat barely 50 per cent of the organic waste generated.[22] The average collection efficiency of municipal solid waste in India is relatively low, ranging from 22–60 per cent.[23]

Construction and demolition waste in India comprises concrete, plaster, bricks, metal, wood, plastics, etc. It is estimated that industrial construction in India generates about 10–12 million tons of waste annually, and nearly 50 per cent of this waste is not currently recycled.[24]

It is estimated that the plastic consumption in India was 8 million tons in 2008, of which about 5.7 million tons of plastics are converted into waste annually.[25] It has been reported that 60 per cent of the total plastic

---

[20]   Ibid.
[21]   Dimpal Vij, 'Urbanization and solid waste management in India: Present practices and future challenges' (2012) 37 Procedia – Social and Behavioral Sciences 437.
[22]   Ministry of Environment and Forests, 'Report' (n 17).
[23]   Ibid.
[24]   Ibid.
[25]   Ibid.

waste generated is recycled but 40 per cent is treated as litter and remains uncollected. Therefore, every day, approximately 6289 tons of plastics are neither collected nor recycled.

The amount of e-waste (including imported e-waste) present in India in the year 2005 has been estimated at 146080 tons, and it is expected to exceed 800000 tons by 2012.[26] It is reported that about 95 per cent of e-waste is processed by the informal sector in India.[27] In order to address this problem, around 23 recycling facilities are currently operating, which when fully operational could recycle 60 per cent of the estimated annual e-waste inventory.[28]

9.2.3 Japan

The rapid development of Japan's post-war economy was at the cost of raw material consumption, high energy costs and heavy environmental pollution. In the past few decades, the rapid growth of the economy has generated a huge volume of multiple solid waste and industrial waste. After the 'energy crisis' of the 1970s, the Japanese gradually realized the importance of solid waste disposal and the government of Japan began to develop the technology for reuse, recycling and effective use of solid waste.[29] The total amount of multiple solid wastes generated in Tokyo was 4.9 million tons in 1989, however this decreased by 26.5 per cent to 3.5 million tons in 1999, mainly because of a series of solid waste management measures taken by the government.[30]

General waste and industrial waste are the two major parts of solid waste generated in Japan.[31] After 1990, the solid waste generation

---

[26] Ministry of Environment and Forests, 'Guidelines for Environmentally Sound Management of E-Waste' (2008) available at <www.moef.nic.in/sites/default/files/guidelines-e-waste.pdf> (last accessed on 11 August 2015).

[27] Maheshwar Dwivedy and R.K. Mittal, 'Willingness of residents to participate in e-waste recycling in India' (2013) 6 Environmental Development 48.

[28] Amit Jain, 'E-waste management in India: Current status, emerging drivers and challenges', Regional Workshop on E-waste/WEEE Management (8 July 2010) Osaka, Japan, available at <http://gec.jp/gec/jp/Activities/ietc/fy2010/e-waste/ew_1-2.pdf> (last accessed on 11 August 2015).

[29] Zhenhua Liu and Yi-ling Guo, 'Current situation of treatment and reuse of solid waste in Japan' (2003) 4 Journal of Qingdao Institute of Architecture and Engineering 87 (in Chinese).

[30] Lin Chen, 'Japan Tokyo waste management experience and enlightenment' (2010) 1 Urban Management Science & Technology 74 (in Chinese).

[31] Wenxin Jian, 'Management and disposal technology of solid waste in Japan' (2002) 4 Environmental Science Trends 1 (in Chinese).

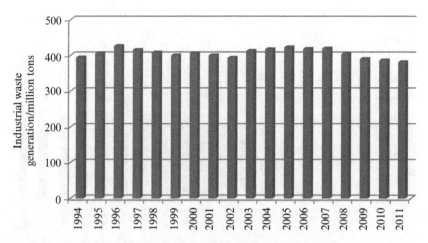

Figure 9.4   Industrial waste generation in Japan, 1994–2011

followed a slightly increasing trend, however since 2000 it has shown signs of decreasing.

The total amount of industrial waste generated in Japan from 1994 to 2011 is shown in Figure 9.4. We can see that, unlike other countries in Asia, the industrial waste generated in Japan in these years is maintained at a relatively stable level, about 400 million tons. Figure 9.5 shows the amount of industrial waste treated in Japan;[32] the final disposal quantities of waste decreased significantly from 69 million tons in 1994 to 12 million tons in 2011, while the recycling and reduction volume increased in the last few decades.

General waste in Japan mainly refers to household waste, including some wastes generated by shops, factories and offices.[33] In the 1980s, the rapid growth of the economy and the great improvement in people's living standards resulted in a sharp increase of waste production. The amount of general waste generated gradually increased from 49.9 million tons in 1989 to 54.83 million tons in 2000. The huge amount of household waste was mainly due to the pursuit of convenience goods, such as durable consumer materials (e.g. household electric appliances, furniture, automobiles,

---

[32]   Ministry of the Environment, 'The discharge of industrial wastes and processing condition in Japan' (2011) (in Japanese) available at <www.env.go.jp/recycle/waste_tech/ippan/h25/data/env_press.pdf> (last accessed on 12 August 2015).

[33]   Wenxin Jian (n 31).

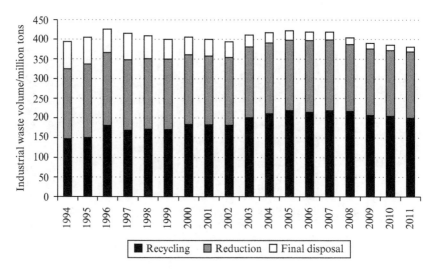

Figure 9.5   Treatment of industrial waste in Japan, 1994–2011

plastic bags, packaging paper, etc.).[34] In the twenty-first century, due to the decline in the population and an increasing awareness of environmental protection, the amount of general waste began to decrease after reaching a peak in 2000. The amount of general waste generated in Japan from 1994 to 2011 is shown in Figure 9.6.[35]

The three main disposal methods of general waste generated in Japan are landfill, incineration and reclamation. Due to the limited land mass and high population density, incineration has become the most important way to solve the waste problem. Incineration accounts for more than 70 per cent of waste treatment, and Japan also has a strong advantage in incineration processing technology.[36] Japan began disposing MSW by incineration from about 1960, and today, possesses the world's leading MSW incineration facilities. In 2013, there were 1172 incineration facilities in Japan with a disposal capacity of 1182.683 thousand tons per day.[37]

[34]   Zhenhua Liu and Yi-ling Guo (n 29).

[35]   See <www.env.go.jp/recycle/waste_tech/ippan/stats.html> (last accessed on 11 August 2015).

[36]   Ministry of the Environment, 'Solid Waste Management and Recycling Technology of Japan- Toward a Sustainable Society' (2008) available at <www.env.go.jp/en/recycle/smcs/attach/swmrt.pdf> (last accessed on 11 August 2015).

[37]   Ministry of the Environment, 'The discharge of general waste and processing condition in Japan' (2013) (in Japanese) available at <www.env.go.jp/recycle/waste_tech/ippan/h25/data/env_press.pdf> (last accessed on 11 August 2015).

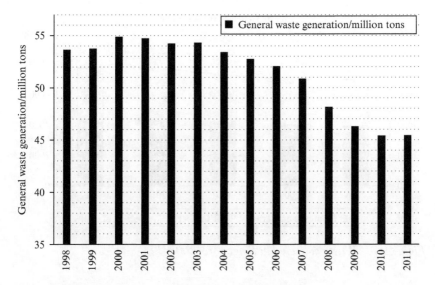

Figure 9.6   General waste generation in Japan, 1994–2011

9.2.4   Singapore

The consumption of resources inevitably leads to the production of waste. Over the past four decades, as Singapore's economy and population has grown and the amount of solid waste generated has increased significantly. The daily average amount of waste collected from 2008 to 2012 is shown in Figure 9.7.[38] The quantity of waste disposed increased from 1 260 tons per day in 1970 to 8289 tons per day in 2013.[39]

As a small city state with high population density, Singapore's main challenge in solid waste management is limited land for waste disposal. Waste minimization and recycling are key components of solid waste management system. Since the early 1990s, Singapore has been actively promoting waste minimization and recycling.[40] In 2013, the overall rate of recycling reached 61 per cent up from 49 per cent in 2005. In the

---

[38]   National Environment Agency, 'Environmental Protection Division Annual 2012 Report', available at <www.nea.gov.sg/docs/default-source/corporate/annual-report/epd-annual-report-2012.pdf?sfvrsn=0> (last accessed on 14 August 2015).

[39]   See <www.nea.gov.sg/energy-waste/waste-management> (last accessed on 14 August 2015).

[40]   Ibid.

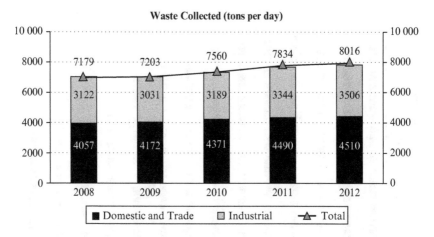

Figure 9.7   Waste collected in Singapore, 2008–12

'Sustainable Singapore' Blueprint, Singapore has set recycling targets of 65 per cent by 2020 and 70 per cent by 2030.[41]

In the 1960s and 1970s, Singapore relied on a number of landfills around the island to handle the solid waste generated on the island. However, in the late 1970s, space constraints made it clear that an alternative method of solid waste disposal was needed. As waste-to-energy (WTE) incineration can reduce waste volume by 90 per cent, Singapore adopted the first WTE plant in 1979. Today, the solid waste disposal facilities comprise four waste-to-energy plants, namely the Tuas Incineration Plant ('TIP'), the Tuas South Incineration Plant ('TSIP'), the Senoko Waste-to-Energy Plant ('SWEP'), and the Keppel Seghers Tuas ('KST'); as well as two landfills: the Semakau Landfill and the Tuas Marine Transfer Station.[42] The capacities of the WTE plants are shown in the Table 9.3 below.[43] The total effective incineration capacity of the four existing waste-to-energy plants is 7600 tons per day.[44]

In 2013, a total amount of 3.03 million tons of waste was disposed of at

[41]   Ibid.

[42]   See <www.nea.gov.sg/energy-waste/waste-management/solid-waste-manage ment-infrastructure> (last accessed on 14 August 2015).

[43]   National Environment Agency, 'Environmental Protection Division Annual 2012 Report' (n 38).

[44]   Ibid.

Table 9.2   Types and amounts of waste disposed of and recycled in Singapore, 2013

Waste Type	Waste Disposed of (tons)	Waste Recycled (tons)	Waste Generated (tons)	Recycling Rate (%)
Construction debris	12 300	1 683 000	1 695 300	99
Used slag	8900	344 800	353 700	97
Ferrous metals	46 800	1 369 200	1 416 000	97
Scrap tyres	2600	19 000	21 600	88
Non-ferrous metals	21 100	114 000	135 100	84
Wood	77 800	254 600*	332 400	77
Paper/Cardboard	581 700	679 400	1 261 100	54
Horticultural waste	131 700	120 900*	252 600	48
Glass	58 900	14 600	73 500	20
Food	696 000	100 000	796 000	13
Plastics	741 100	91 100	832 200	11
Textile/Leather	140 300	16 300	15 600	10
Ash and Sludge	176 400	14 200	190 600	7
Others (stones, ceramics, rubber etc.)	330 000	4800	334 800	1
Total	3 025 600	4 825 900	7 851 500	61

Note: *Includes 131 900 tons used as fuel in biomass plants.

Source: <www.nea.gov.sg/energy-waste/waste-management/waste-statistics-and-overall-recycling> (last accessed on 14 August 2015).

Table 9.3   Capacities of the WTE plants

Facility	TIP	SWTE	TSIP	KSTP
Ownership	Govt	Keppel	Govt	Keppel
Year commissioned	1986	1992	2000	2009
Capacity (tons/day)	1700	2100	3000	800
% capacity	22%	28%	39%	11%

the refuse disposal facilities. Of this total, approximately 2.83 million tons or 92 per cent were incinerated while the remaining 0.20 million tons were landfilled. Table 9.4 shows the amount of solid waste disposed of at the disposal sites from 1999 to 2013.

Waste management and the Green Economy

*Table 9.4   Amount of solid waste disposed of at the disposal sites,
              1999–2013*

Year	Landfilled (Thousand Tons)	Incineration Plants (Thousand Tons)	Total Refuse Disposed of (Thousand Tons)
1999	756.2	2036.30	2792.50
2000	357	2440.20	2797.20
2001	251.3	2550.90	2802.20
2002	204.3	2421.30	2625.60
2003	193.8	2311.20	2505.00
2004	219.6	2263.00	2482.60
2005	270.1	2278.60	2548.70
2006	234.5	2329.10	2563.60
2007	187.3	2379.50	2566.80
2008	177.8	2449.80	2627.60
2009	148.9	2480.00	2628.90
2010	174.1	2585.40	2759.50
2011	203.5	2656.00	2859.50
2012	198.0	2736.00	2933.90
2013	200.4	2825.10	3025.50

9.2.5   Republic of Korea

In order to improve the waste management capacity in the Republic of Korea, the 'Waste Control Act' was adopted to gather the necessary baseline data on waste generation and treatment.[45] The amount of waste generated nationwide by type, regional distribution, and changes in disposal patterns is reported in the annual environmental statistics yearbook.

According to the environmental statistics yearbook 2012 of the Republic of Korea, the total amount of waste generated in the Republic of Korea has gradually increased from 261 032 tons per day in 2001 to 374 642 tons per day in 2010, but domestic waste disposed of per person has decreased from 1.04 kg per day in 2002 to 0.96 kg per day in 2010.[46] In particular, the amount of landfilled and incinerated wastes

---

[45]   Source: <www.un.org/esa/dsd/dsd_aofw_ni/ni_pdfs/NationalReports/korea/WasteManagement.pdf> (last accessed on 12 August 2015).

[46]   Republic of Korea, 'Environmental statistics yearbook 2012 of Republic of Korea', available at <http://eng.me.go.kr/eng/web/index.do?menuId=29&findDepth=1> (last accessed on 11 August 2015).

have greatly decreased since 2001 due to the continuous increase in recycling following the implementation the Volume-Based Waste Fee System.[47] However, the amount of general industrial waste has increased annually.

The recycling rate of domestic waste in the Republic of Korea has increased and the percentage that is landfilled has decreased, while the percentage incinerated has also increased. In 2001, 43.3 per cent of municipal solid wastes were sent to landfill and 43.1 per cent were recycled, whereas in 2010, 60.5 per cent were recycled and 17.9 per cent were sent to landfill.[48]

General industrial waste has demonstrated a similar pattern to domestic waste. The percentage of general industrial waste that is landfilled has decreased, whereas the percentage recycled has steadily increased, reaching 86.9 per cent in 2010.

9.2.6 Mongolia

Mongolia is a landlocked country with an area of 1.565million km^2, the population is about 2.94 million of which nearly 44 per cent (about 1.3 million) live in the capital city of Ulaanbaatar (UB).[49] Solid waste generation in Mongolia is 2.9 million tons per year, 1.1 million tons of which is generated in UB. The main composition of solid waste in Mongolia is household waste (85 per cent), construction waste (12 per cent), medical waste (0.3 per cent), hazardous waste and chemicals (3 per cent). Due to the high population density and that large quantity of waste in UB, solid waste management and disposal is a severe problem for the local government. There are about 450 waste collection points (open sites), which cover over 3000 hectares of land. Of these 450 sites, about 220 are in UB, but the collection of waste is not well organized.[50] It was reported that there were three centralized landfill sites in UB in 2005, which were unable to cope with all the waste generated daily. In 2009, about 2500 tons

---

[47]   See<http://eng.me.go.kr/eng/web/index.do?menuId=139&findDepth=1>(last accessed on 11 August 2015).

[48]   See<http://eng.me.go.kr/eng/web/index.do?menuId=140&findDepth=1>(last accessed on 11 August 2015).

[49]   See <www.fmprc.gov.cn/mfa_chn/gjhdq_603914/gj_603916/yz_603918/1206_604450/> (in Chinese) (last accessed on 11 August 2015).

[50]   World Health Organization, 'Environmental Health Country Profile – Mongolia' (14 February 2005) available at <www.wpro.who.int/rfeh/country_profiles/mongolia.pdf?ua=1> (last accessed on 12 August 2015).

of solid wastes were sent to landfill per day in UB, while 300 tons of solid wastes were dumped.[51]

9.2.7   Pakistan

Pakistan has a population of 197 million, with over 35 per cent people living in urban areas. Like many other developing countries in Asia, there is no statistical data or reliable research on national waste generation. According to the website of the Environment Protection Department of Pakistan, solid waste generation in Pakistan ranges between 0.283 to 0.612 kg/capita/day and the waste generation growth rate is 2.4% per year.[52] There are many problems regarding solid waste management in Pakistan: there is no proper waste collection system; waste is dumped on the streets; different types of waste are not collected separately; and there are no controlled sanitary landfill sites, etc. It is reported that only 60 per cent of the waste generated is actually collected in most Pakistani cities and more than 90 per cent of this is disposed through open dumping.[53] Solid domestic waste in Pakistan is typically dumped on low-lying land.[54]

9.3   ISSUES AND CHALLENGES OF WASTE MANAGEMENT AND DISPOSAL IN ASIA

9.3.1   Lack of Energy and Resources

Global resource consumption in Asia is increasing rapidly, and the material use has gone up by eight times in the last century.[55] Based on a study by the UNEP, global material extraction, including biomass, construction minerals, fossil energy carriers, and ores and industrial

---

[51]   Ibid.

[52]   See <www.fmprc.gov.cn/mfa_chn/gjhdq_603914/gj_603916/yz_603918/1206_604018/> (in Chinese) (last accessed on 11 August 2015).

[53]   S. A. Batool and M. N. Ch, 'Municipal solid waste management in Lahore city district, Pakistan' (2009) 29(6) Waste Management 1971.

[54]   See <www.fmprc.gov.cn/mfa_chn/gjhdq_603914/gj_603916/yz_603918/1206_604018/> (in Chinese) (last accessed on 11 August 2015).

[55]   Fridolin Krausmann, Simone Gingrich, Nina Eisenmenger et al, 'Growth in global materials use, GDP and population during the 20th century' (2009) 68(10) Ecological Economics 2696, available at <www.uni-klu.ac.at/socec/down loads/2009_KrausmannGingrichEisenmenger_Growth_in_global_materials_use_EE68_8.pdf> (last accessed on 14 August 2015).

minerals, is increasing steadily, associated with the international GDP growth.[56] Annual extraction of ores, minerals, hydrocarbons and biomass has grown from 7 billion tons in 1900 to 60 billion tons in 2014 and, on current trends of growth in population and economic activity, is set to reach 140 billion tons by 2050.[57] At the same time, the resource consumption in the Asia-Pacific region is also on the rise, while the overall resource efficiency has remained poor especially in some developing countries.

Taking the case of China, it is well known that China has a large amount of natural resources, but divided by the head count, this amount becomes smaller. Mineral resources play an important role in China's economic and social development. More than 95 per cent of energy and 80 per cent of industrial raw materials come from mineral resources.[58] China is also the largest energy producer and consumer in the world. The annual energy consumption in China is increasing year by year, from 1504.06 million tons of standard coal in 2001 to 3617.32 million tons of standard coal in 2012.[59] At present, China mainly relies on the use of non-renewable resources such as raw coal, crude and other non-renewable energy. In 2012, the proportion of consumption of coal and fossil oil in China was 85.4 per cent,[60] far higher than the world average proportion of consumption.[61] The renewable energy utilization efficiency is still very low in China, in 2012, the proportion of consumption of natural gas and hydropower, nuclear and wind power was 14.6 per cent.[62]

---

[56]   UNEP, 'Towards a Green Economy: Pathways to Sustainable Development and Poverty Eradication' (UNEP 2011) available at <www.unep.org/greenecon omy/Portals/88/documents/ger/ger_final_dec_2011/Green%20EconomyReport_ Final_Dec2011.pdf> (last accessed on 11 August 2015).

[57]   UNEP, 'Management and conserving the nature resource best for sustained economic and social development' (2014) available at <www.unep.org/resource panel/Portals/24102/IRP%20Think%20Piece%20Contributing%20to%20the%20 SDGs%20Process.pdf> (last accessed on 11 August 2015).

[58]   Song Xinyu, 'Current status and sustainable development strategy of mineral resources in China' (1997) 16(59) Exploration of Nature 27 (in Chinese).

[59]   See <www.stats.gov.cn/tjsj/ndsj/2013/indexch.htm> (in Chinese) (last accessed on 11 August 2015).

[60]   Ibid.

[61]   British Petroleum, 'Statistical Review of World Energy 2013', available at <www.bp.com/content/dam/bp/pdf/statistical-review/statistical_review_of_ world_energy_2013.pdf> (last accessed on 12 August 2015).

[62]   See <www.stats.gov.cn/tjsj/ndsj/2013/indexch.htm> (in Chinese) (last accessed on 11 August).

9.3.2   Huge Amount of Solid Waste Generation and Future Complex Waste Types

The growth of the economy and the percentage of the urban population have led to a change of lifestyle and consumption levels in many countries, which has directly resulted in the growth of the quantity as well as the changing characteristics of the waste generated. It is estimated that more than 1.3 billion tons of municipal solid wastes were generated in 2012 and that some 2.2 billion tons a year will be generated by 2025 in the whole world.[63] In India, the volume of waste is on the rise, as economic growth drives more and more people from the rural hinterland to the urban areas, spawning new consumption patterns and social linkages. The urban population generated about 114 576 tons of MSW per day in 1995. The figure is predicted to reach 440 460 tons per day in 2026 based on the rapid growth of the population and the economy. The large metropolises of India now generate more than 6000 tons of solid waste per day, and Delhi alone generates more than 3500 tons.[64] By 2030, India will probably generate more than 125 000 metric tons of waste every year.[65] The trend is more or less similar in countries such as Bangladesh, Afghanistan, Pakistan, Nepal, and Sri Lanka.

Although the level of industrial cleaner production in China has increased year by year, and the industrial solid waste generation per dollar of GDP produced is decreasing, the accelerated economic growth in the '12th Five-Year Plan'[66] period results in a rapid improvement of the national economy and industry. Based on this situation, it is expected that, in the next ten to 20 years, the industrial solid waste generated in China will continue to show a substantial growth trend. It is predicted that, with the increase of the urban population and economic growth, the national MSW and industrial solid waste generation in China in the future will show a rapid growth trend, with an average annual growth rate of about 2.4 and 6.5 per cent respectively. Based on the statistical data of China

---

[63]   Daniel Hoornweg and Perinaz Bhada-Tata, 'What a Waste: A Global Review of Solid Waste Management', Urban Development Series Knowledge Papers (World Bank 2012) available at <http://siteresources.worldbank.org/INTURBANDEVELOPMENT/Resources/336387-334852610766/What_a_Waste2012_Final.pdf> (last accessed on 3 August 2015).

[64]   Ray (n 2).

[65]   Ibid.

[66]   National People's Congress of the People's Republic of China, '12th Five-Year Plan for national economic and social development in China (2010–2015)' (in Chinese) available at <www.npc.gov.cn/wxzl/gongbao/2001–03/19/content_5134505.htm> (last accessed on 1 September 2015).

today, the combined amount of MSW and industrial solid waste generation will reach 416.69 million tons in 2020 and 10431.25 million tons in 2030. Associated with the development of society, the emerging waste streams, such as e-waste, packing waste etc., are expected to increase. The waste type will be more complicated in the future, which could bring more pressure to resource utilization and solid waste pollution prevention in Asia.

9.3.3   Poor Implementation of Waste Classification and Recycling

Recycling has been widely accepted as a sustainable solid waste management method because of its potential to reduce disposal costs and waste transport costs, and to prolong the life spans of landfill sites. Resource recovery rates of solid waste in Asia are relatively low, and the overall development of waste recycling is not balanced. Some developed countries in Asia, such as Japan, the Republic of Korea, and Singapore, have been actively promoting waste reduction and recycling over the last few decades, and the recycling rate of waste has significantly increased.

In Japan, the general waste has been classified scientifically, so the collected waste can mostly be recycled. In 2013, the overall rate of recycling reached 61 per cent in Singapore up from 49 per cent in 2005, while the recycling rate of some waste types was over 90 per cent. To promote recycling, the government of the Republic of Korea has been administering an Extended Producer Responsibility (EPR) system since 2003, and since the introduction of this system, the total amount of waste generated per person has increased by 14.0 per cent, from 46.62 kg in 2003 to 53.16 kg in 2007, while the amount recycled increased by 30.5 per cent, from 21.88 kg in 2003 to 28.56 kg in 2007.[67] In other developed countries in Asia, like Malaysia, the current recycling rate is about 5 per cent, though Malaysia set the objective of 22 per cent of total solid waste being recycled by the year 2020.[68] Implementation of waste classification and recycling in developing countries in Asia, such as India and China, is very poor. In India, all waste – whether it is biodegradable, recyclable, construction, hazardous or solid – is mixed together. While the collection efficiency is 60 per cent, the remaining 40 per cent lies uncollected and scattered all over the towns and cities, polluting the surrounding land and water resources.[69] In China,

---

[67]   Republic of Korea, 'Integrated Waste Management Plan' (n 45).
[68]   R. P. Singh, P. Singh, A. S. F. Araujo, et al, 'Management of urban solid waste: Vermicomposting a sustainable option' (2011) 55(7) Resources, Conservation and Recycling 719.
[69]   Ministry of Environment and Forests, 'Report' (n 17).

numerous MSW containers with recyclable and non-recyclable signs have been placed in residential and commercial regions to facilitate the separation and recycling of MSW. However, because of insufficient public outreach, most residents cannot distinguish whether items are recyclable or non-recyclable and still randomly discard waste.

9.3.4   Lack of Disposal Facilities for Solid Waste

Treatment and disposal technology is backward in Asia, and the solid waste disposal level is relatively low. In China, incineration and landfill are the main ways of treatment of harmless MSW, but the incineration proportion is still very low. In 2012, the proportion of MSW incineration was 25 per cent, while landfill was 72 per cent.[70] In India, there are no specifically designed landfill sites in class II and class III cities to dump the waste. Equipment used for collection and transportation of waste is very old, and the only method to recycle the waste is incineration which creates serious health and environmental hazards when all mixed waste is burned.[71] It is reported that in 2008 there were about 24 landfill facilities in India, jointly having the capacity of holding 0.06 MMT/d of waste, while the total requirement for land filling was about 0.183 MMT/d.[72]

9.4   OPPORTUNITIES

9.4.1   Great Market Potential for Energy and Resource Recovery from Waste

Waste is not something to be abandoned or discarded, but rather is a valuable resource. An example is electrical and electronic waste (e-waste). One ton of e-waste contains as much gold as five to 15 tons of typical gold ore, and amounts of copper, aluminium and rare metals that exceed by many times the levels found in typical ores. As the main components of MSW, waste rubber, plastic, paper and glass are all recyclable resources. In China, these recyclable resources are called 'urban mining'. If a sound and proper method is used, waste management can deliver several benefits. According to the report by UNEP,[73] the global waste market, from

---

[70]   See <www.stats.gov.cn/tjsj/ndsj/2013/indexch.htm> (in Chinese) (last accessed on 3 August 2015).

[71]   Vij, 'Urbanization and solid waste management in India' (n 21).

[72]   Ministry of Environment and Forests, 'Report' (n 17).

[73]   UNEP, 'Towards a Green Economy' (n 56).

collection to recycling, is estimated at US$410 billion a year, not including the sizable informal segment in developing countries. When efficient practices are introduced into production and consumption, valuable materials are recovered. Through waste reduction and recycling, the adverse influence on the environment caused by improper disposal of waste will be reduced.

Secondary material markets, e.g. for metals, recovered cellulose fibre and paper, play an important role in minimizing resource consumption and increasing waste utilization on a global basis. Asia makes up the most dynamic and arguably the most important recycling market. According to a report released by the National Development and Reform Commission of the PRC,[74] in 2011 the quantity of main recyclables collected, including iron and steel scrap, nonferrous metal, plastics, tyres, paper, e-waste, scrapped car and ships, reached 165 million tons, which is twice the amount collected in 2005. The total value reached 576.39 billion RMB, 12.7 per cent higher than that of 2010.

9.4.2 Large-scale Industry will be Formed due to the Enormous Quantity of Waste

Waste generation is increasing with the rapid population and economic growth. It is estimated that in 2012, the whole world produced more than 1.3 billion tons of MSW, and the amount is expected to be 2.2 billion tons by 2025.[75] The enormous quantity of waste leads to a great demand for a waste treatment industry. The solid waste treatment industry in Asia is however still at an early stage of development, and the degree of industrialization and market concentration is very low. The competitive pattern has not been established, and the competition in the market is in a state of disorder. But more and more countries are focusing on waste management and recycling, and have launched a series of measures to promote the development of the solid waste treatment industry. The rise of the industry is projected to accelerate in the coming decades, perhaps sharply. It is estimated that by 2030, the amount of MSW generated in China will increase to about 480 million tons, while by 2020, the investment in solid waste treatment might reach 681.4 billion Yuan, accounting for 3 per cent of the total investment of environmental protection.

---

[74] National Development and Reform Commission, 'Chinese annual reports of resources comprehensive utilization (2012)' (2013) (in Chinese) available at <www.ndrc.gov.cn/fzgggz/hjbh/hjzhdt/201304/t20130412_536838.html> (last accessed on 3 August 2015).

[75] Hoornweg and Bhada-Tata, 'What a Waste' (n 63).

9.4.3   Great Requirements for Advanced Technology and Complete Equipment

China is the largest energy consumption country in the world. It is estimated that in 2020, China will face an energy gap of 0.5–1 billion tons of standard coal. And the air pollution situation in China is grim, especially in the capital Beijing. According to a government report on sources of air pollution in Beijing City, coal burning accounts for 22.4 per cent. Total greenhouse gas emissions in China have surpassed those of the US, and have been ranked first in the world since 2007. In order to reduce carbon dioxide emissions, it is necessary to improve the proportion of non-fossil energy consumption. Energy produced from organic waste has the characteristics of being energy rich, renewable, clean, environmentally protective, and has zero emissions of carbon dioxide. China has the largest amount of organic waste; according to a report, the estimated amount of bio-based resources is 3.5 times greater than water resources, and two times greater than wind resources. Statistics show that domestic demand for MSW incineration power generation equipment will reach US$ 5 billion over the next five years. The requirements of advanced MSW incineration technologies, in addition to the necessary equipment and services for the progress of MSW to energy industries in China, will provide exciting opportunities for investors all over the world. The energy conservation law of the PRC encourages the introduction of advanced energy conservation technology and equipment from abroad. Pakistan,[76] being an agricultural country, is rich in biomass energy sources, and also has a great demand for advanced technology and complete equipment.

9.4.4   Huge Demand for Technical Assistance and Engineering Services

Compared with other developed countries, the work of waste management and recycling in most Asian countries started late. In order to solve the existing problems caused by solid waste as soon as possible, developing countries in Asia need to benefit from the experience of advanced countries and regions both within and outside Asia, and continuously improve the level of solid waste management. The technology for processing solid waste in developing Asian countries currently is relatively

---

[76]   M. K. Farooq and S. Kumar, 'An assessment of renewable energy potential for electricity generation in Pakistan' (2013) 20 Renewable and Sustainable Energy Reviews 240.

backward. Ensuring the availability and functioning of technology and equipment will require significant time and effort. Accordingly, technical assistance and engineering services from highly advanced countries and regions is the best way to rapidly improve the level of treatment of solid waste in Asia.

Conclusions

Katharina Kummer Peiry, Andreas R. Ziegler and Jorun Baumgartner

---

Does resource and energy recovery from waste have the potential to become a pilot area for a Green Economy? As noted in the Introduction, this question inspired the collection of essays in this book. The contributions offer a somewhat kaleidoscopic outlook, providing a range of diverse but complementary insights from different angles and perspectives. Not surprisingly, it is difficult to draw a clear-cut answer from them. They do however add up to a range of elements that may be linked together to form the basis of an answer.

Part I of the book focuses on the role of international law and policy in shaping the approach to waste management, including resource and energy recovery from wastes. An overview of the general principles of international law as they relate to waste management is followed by an examination of whether and how international law supports a resource-based approach to waste. The role of the Basel Convention as the sole global treaty addressing waste management is then considered. Finally, waste as potential tradable goods under the WTO agreements is analysed.

Rosemary Rayfuse's contribution in Chapter 1 shows that the issues arising in the context of waste management have been present in international law for a long time. Through the examination of the interpretation and application of the general principles of international law that are particularly relevant in the context of waste management (the principles of permanent sovereignty over natural resources; of preventive action; of cooperation; of sustainable development; and the precautionary principle), and the more recent principles aimed specifically at waste management (self-sufficiency, proximity of disposal, waste minimization, environmentally sound management, and prior informed consent to waste imports), she shows that international law *has* developed in response to environmental issues and necessities, and more importantly, has the *capability* to develop in the face of important environmental challenges. Yet, the author concludes that principles alone do not solve the challenges, but rather provide interpretative guidance on how the law has been developed and how it should continue to develop in the future.

Tarcísio Hardman Reis foreshadows in Chapter 2 the cross-cutting nature of waste management. By conceptualizing the treatment of wastes under international law from three different angles, namely human rights, environmental protection, and economic resources, he offers an unusual and holistic approach to waste management that would be visibly necessary if one were to try and capture the different policy areas and challenges involved in modern waste management in one single legal instrument. He identifies problems of application of existing legal frameworks stemming in particular from the absence of a uniform, all-encompassing definition of wastes, and the fact that the distinction between hazardous and non-hazardous wastes creates a (sometimes tricky) tension between the principle of *control* of hazardous wastes on the one hand and the *free trade* in non-hazardous wastes on the other – a theme central to the WTO context discussed in Chapter 6. The existing gaps are increasingly filled by 'soft law' instruments (principles, concepts and technical standards), which in his opinion make an important contribution towards perceiving waste as a resource.

Chapters 3 and 4 offer two complementary, though somewhat contrasting views on the Basel Convention and its political impact. In Chapter 3, *Pierre Portas* draws a historical-political picture of the Basel Convention, providing an insider's insight into the factors that gave rise to the negotiations of the Basel Convention in the 1980s, and the developments that have shaped the Convention since then. He remains however sceptical of the Convention's role to promote waste management as part of a Green Economy. Central to this assessment is the tension between environmental protection and free trade, with free trade in his view continuing to be given priority, as well as the continuing lack of capacity of many States Parties to recycle hazardous wastes in an environmentally sound manner.

In her review of the recent political development of the Convention, *Juliette Voïnov Kohler* in Chapter 4 strikes a more positive note with respect to the Convention's potential to serve as a basis for moving towards a resource-oriented approach. The fact that States managed to overcome, at the 10th Conference of the Parties in 2011, the long-term political deadlock and achieved political consensus on the Ban Amendment and on the necessity to strengthen the Convention by putting more emphasis on the 'prevention, minimization and reduction of wastes' aspect, shows in her view the contribution the Basel Convention can make to 'promoting sustainable livelihoods', but also to recognizing waste as a resource.

The treatment of waste as an economic resource – as tradable goods – under WTO law is the focus of *Mirina Grosz*'s contribution in Chapter 5. Her assessment of the potential incompatibility of waste movement restrictions imposed by the Basel Convention and other instruments with WTO law shows that WTO jurisprudence still grapples with the dichotomy of trade versus non-trade concerns. However, she finds that restrictions to cross-border movements of *hazardous* wastes are more likely to be justified when implemented with a view to environmental and human health concerns, while this remains more uncertain for *non-hazardous* wastes, left largely unregulated and uncontrolled.

Part II of the publication takes the discussion from the legal and policy level to the concrete, delving into more practical aspects as regards the opportunities and challenges lying in a Green Economy approach towards waste management. Following an introduction to the concept of a Green Economy, several concrete examples are given of how this can operate, in areas diverse as waste electrical and electronic equipment and landfill-to-gas. The significance of turning wastes into resources for Asia – probably the part of the world for which this is the most relevant at this time and in the years to come – is also presented.

In Chapter 6, *Vera Weick* gives a comprehensive overview of the emergence of the concept of 'greening the economy' as a tool to promote sustainable development, and explains some of the reasons why sustainable development has been on the international agenda for decades, but has not been satisfactorily implemented. Despite the rhetoric, there is still a 'grow first, clean up later' approach in development policies, and practical implications of sustainable development have to date remained limited and its lines sketchy. The Green Economy concept is an attempt to remedy this, and related concepts developed by different actors for their focus activities, such as the concept of a Circular Economy, may be most directly relevant to addressing waste and resource management. Yet, there is no 'one size fits all' concept and measures to facilitate transition to a Green Economy must be tailored to each country's specific circumstances.

Chapters 7–9 then turn to select methods and technologies of resource and energy recovery, and challenges and opportunities arising therefrom.

In Chapter 7, *Mathias Schluep* examines the challenges and opportunities arising from waste electrical and electronic equipment (WEEE). As the author explains, the perception of WEEE has been shifting from one of problem to one of opportunity. This is mainly due to the fact that many of these waste appliances contain valuable metals and/or minerals (also called 'urban minerals'), and actors have thus become increasingly aware of the economic interest in recovering these materials. Yet, this *economic* opportunity only becomes an *environmental* opportunity if these materials – often highly toxic ones – are recycled and recovered in an environmentally sound manner. The author discusses the presently applied methods and technologies to recycle WEEE both from a developed country and from a developing country perspective and finds that challenges remain for both. Introducing the notion of recycling efficiency, he shows that both formal and informal recycling systems have their strengths and weaknesses, and that overall recovery efficiency may be diminished by poor collection and pre-processing efficiency. He concludes that an international division of labour in WEEE recycling could link geographically distributed treatment options, combining pre-treatment at the local level and end-processing in state-of-the art facilities in industrialized countries; and that efficient and sustainable recovery as a raw material is a market opportunity – though one that will require harmonization of international standards and processes to identify 'fair' secondary resources if they are to be leveraged, as well as the use of international financial mechanisms for those materials covered by such mechanisms.

Jessica North, in Chapter 8, explores the challenges and opportunities of sound waste management for the purpose of energy generation and greenhouse gas reduction. As the author explains, landfill gas (LFG) is a significant contributor to greenhouse gas emissions; yet, if properly extracted and combusted in a power generation facility, LFG could also be a potential source of 'green' power, thus providing an opportunity to 'kill two birds with one stone'. However, even though the technology exists, technical, political and financial challenges remain (in particular, but not only, for developing countries), leaving a large untapped potential. The author gives an overview of how LFG recovery and energy generation works, and draws for this purpose on experience of the Australian LFG industry.

In Chapter 9, finally, *Jinhui Li*, *Xiaofei Sun* and *Baoli Zhu* take a closer look at the challenges and opportunities for economically and environmentally sound energy and resource recovery in Asia. As the authors note, waste management and resource recovery is a vital challenge in Asia, where more than half of the world population lives and which has seen skyrocketing economic growth in particular over the past two decades. Yet, waste management in many Asian countries is still based on outdated technologies. The authors present an overview of the current situation and challenges in a range of Asian countries, both developing and developed. They find that many of the countries surveyed face similar challenges in terms of efficient and sound waste management, such as the increase in waste generation due to heightened levels of global resource consumption; differing waste treatment capacities depending on the level of economic development of the countries concerned; and in general rather poor resource-use efficiency in developing countries. Based on this analysis, the authors find that resource recovery from waste could present significant opportunities, with countries like China and India facing a particularly increased need for energy in the next decades, which could be met by waste-to-energy schemes. This will however require advanced technology and investment, and relevant support from developed countries within and outside Asia.

What, then, are the conclusions to be drawn from this collection of essays? If one were to formulate an overall conclusion in the briefest of terms, it might be that there is definitely a potential for turning wastes into valuable resources and thus contributing to a Green Economy, but that there are many obstacles to be overcome for this potential to be realized.

Considering the law, we note that its progress in capturing the new paradigm of waste-to-resource, which would appear to be a prerequisite for positioning the waste-to-resource approach as part of the Green Economy, has been rather slow to date. This is hardly surprising, given that by its very nature, the law is slow in encompassing new ideas and approaches. The contributions show that while a basis exists, many questions remain for which the law has yet to find solutions. Most importantly, the lack of an overarching legal framework for all types of wastes and applicable to all stages of its treatment leaves multiple gaps and presents difficulties in the application of existing international instruments.[1] The efficiency of existing international legal frameworks is compromised by the lack of uniform definitions of the different types of waste materials and the piecemeal approach towards the notion of wastes, distinguishing between hazardous and non-hazardous wastes and positioning the treatment of wastes as seemingly contradictory values (trade versus environment), depending on which legal framework is being applied. Further, existing international instruments do not comprehensively capture the process from production to final disposal, including re-entry of re-used or recycled goods into the market, therefore failing to reflect the 'life-cycle approach' towards waste management.

Challenges to a Green Economy approach to waste management also lie in the practical and technical difficulties identified by the authors. A major obstacle to fully exploring the economic potential of waste-to-energy technologies is the lack of financial means and of technical capacities, particularly in developing countries. These factors are also often to blame for the lack of processing capacities for valuable waste materials, such as WEEE, in developing countries. This too does not come as a surprise: the argument is made in every international environmental negotiation that acceptance of obligations by developing countries is subject to provision of resources

---

[1]   See also Katharina Kummer Peiry, 'The Chemicals and Waste Regime as a Basis for a Comprehensive International Framework on Sustainable Management of Potentially Hazardous Materials?' (2014) 23(2) *Review of European Community & International Environmental Law* 174.

to build the necessary capacities. Appeals for more financial support to developing countries for environmental protection measures have been made in every international negotiation process over the past decades. More recently, recognition that public funding is insufficient to build the necessary capacities for environmental protection has led to policy calls for increased involvement of the private sector. A greater involvement and role of the private sector is indispensable, also in light of tighter government budgets as a consequence of economic and fiscal crises. The proposition underlying the notion of a Green Economy in general, and the hypothesis presented in the Introduction to this book in particular, is that investments in environmentally sound management can be made attractive for industry, thus generating the required funds. However, the picture that emerges from the contributions on WEEE and landfill-to-gas, as well as the general overview of the situation in Asian countries, is a fairly sombre one. Commercial investment in these operations is currently not sufficiently attractive. The inconsistencies in legal frameworks, and the uneven implementation of legislation that does exist, contribute to the lack of a level-playing field for potential investors. *Jessica North* aptly sums up the main obstacles, both legal and practical: technical limitations in poorer parts of the world, weak regulatory environments, and lack of financial incentives.

On the positive side, there is political momentum to take action to achieve a more sustainable future, as can be seen from the outcomes of the Rio+20 Summit, the adoption of the SDGs, and the 2015 Paris Climate Agreement, which is being hailed as 'historical'. Concerning waste management and the achievement of a Circular Economy, the 2011 policy decisions under the Basel Convention are particularly significant, given that the Convention is the sole global legal instrument on waste management. We can also observe a change in attitudes – albeit a slow one – at the grass root level in developed (and increasingly also in developing) countries. Sustainable life styles are mushrooming in developed countries and are heralding a change in consumption patterns crucial for a sustainable approach towards waste management; in some segments of society, the status symbol of ownership (at least of certain items of daily consumption or necessity) is increasingly losing its appeal and is more and more replaced by sharing initiatives. 'Zero waste' policies are thriving and have started to be implemented in cities such as San Francisco, which has already achieved a recycling or reuse rate of half of its generated waste and aims at a 'zero waste' rate by 2020.[2]

---

[2]   Giles Atkinson, Simon Dietz, Eric Neumayer and Matthew Agarwala (eds), *Handbook of Sustainable Development* (Edward Elgar, 2nd ed. 2014) 206.

Another positive development in recent years is that despite the obstacles, the private sector has begun to recognize the business potential that lies in certain 'green' products and initiatives. At the grass-root level, private business initiatives such as second-hand shops, repair cafés and sharing initiatives thrive. Grass-root projects that help poor people turn wastes into resources to generate income are also becoming more widespread in developing countries.[3] Institutional investors also are increasingly turning towards 'responsible investing', for example in the form of 'clean energy' investments.[4] The so-called Breakthrough Energy Coalition, a network of dozens of entrepreneurs, billionaires and/ or philanthropists, has recently pledged to invest massively in innovative technologies aimed at (near) zero carbon emissions, and to support, in particular, those countries that have committed to increasing public research into these technologies.[5] This initiative also shows the increasing importance of collaborative efforts between the public and the private sector. Indeed, in the attainment of the SDGs, both sectors have their role to play. As aptly captured by the UNCTAD World Investment Report 2014, their roles are complementary:

> [t]he private sector cannot supplant the big public sector push needed to move investment in the SDGs in the right direction. But an associated big push in private investment can build on the complementarity and potential synergies in the two sectors to accelerate the pace in realizing the SDGs and meeting crucial targets.[6]

---

[3] See for example the account of a project with waste pickers in India under the Basel Convention's Green Heroes series, available at <www. basel.int> (last visited on 14 January 2016), and the Colombian government's initiative 'Computadores para Educar', available at <www.computadoresparaeducar.gov.co> (last visited 30 January 2016).

[4] The global insurance company Allianz declared recently that it will not invest in carbon emission-intensive industries anymore and will instead increase its investments in wind energy, see <www.spiegel.de/wirtschaft/unternehmen/ allianz-zieht-investitionen-aus-kohleindustrie-ab-a-1064208.html> (last visited on 11 December 2015). See also Mark Halle, 'Tipping Permitted: Green Finance Goes Mainstream', IISD Commentary (3 December 2015) available at <www.iisd.org/ commentary/tipping-permitted-green-finance-goes-mainstream> (last visited on 12 December 2015).

[5] See <http://www.breakthroughenergycoalition.com/en/index.html> (last visited 13 December 2015).

[6] UNCTAD, 'World Investment Report 2014: Investing in the SDGs: An Action Plan' (UNCTAD 2014) 137, available at <http://unctad.org/en/ PublicationsLibrary/wir2014_en.pdf> (last accessed on 24 February 2016).

Against this background, the challenge will be to leverage the emerging policy support for a waste-to-resource and Circular Economy approach to fill the gaps in the existing regulatory frameworks, ensure their effective implementation, and build the required technical and financial capacities – or, in more concrete terms, to further develop the emerging legal and policy frameworks described in this book to address the obstacles that are clearly in evidence.

It will be necessary to strengthen and further develop sustainable and environmentally sound waste management policies that decouple waste generation from economic growth and give priority to waste prevention, followed by waste reduction, recycling and recovery (with landfill being the last option) in line with the waste management hierarchy.[7] A Circular Economy approach, where the 'generation of waste is minimized and any unavoidable waste enters a new cycle at the same or higher level of quality',[8] would best respond to the aim of waste prevention and reduction and take into account the fact that resources are finite and therefore must be made the best possible use of before final disposal, yet require incorporation of the '3Rs' (Reduce, Reuse, Recycle) principle into every stage of the supply chain.[9] Yet, investments only thrive in a climate conducive to investment, which requires States to put in place targeted policies that make private sector investment into the SDGs more attractive, all the while taking certain safeguards in particular with respect to essential infrastructure industries.[10] In particular developing countries, which so far most often lack technical and financial capacities for sustainable waste management, could thus benefit from channelled investment to finance state-of-the-art technology, for example for landfill gas-to-energy installations. This however requires the implementation of an investment-enabling framework by the host country, with sufficiently strong investor protections so as to create an investment-friendly environment, while also ensuring that other areas of public interest, such as environmental protection and social development, are adequately considered.[11]

The key would thus appear to be to create a legal and economic environment conducive to investment in the relevant operations. Policy calls for greater private sector involvement are a step in the right direction, but are

---

[7] Atkinson et al (n 2) 204.
[8] Ibid., 207.
[9] Ibid., 208.
[10] Ibid., 150.
[11] See K. Kummer Peiry, R. Khanna, and V. Sahajwalla, 'Resource and Energy Recovery from Wastes: Perspectives for a Green Economy' (2012) 42(6) *Environmental Policy and Law* 346.

clearly insufficient to achieve the desired result. Indeed, if not underpinned by concrete measures, they risk becoming a mere lip service to sustainable development and related concepts. To address the main obstacles that affect especially the poorer parts of the world – technical limitations, weak regulatory environments, and lack of financial incentives – a stable investment environment is needed. An effective regulatory framework is an important part of this.

There is clearly a need to cast policy commitments into concrete legal frameworks, both at the international and at the national and sub-national levels, which would fill the gaps identified in this book. Key elements of such frameworks would include a clear distinction between waste and non-waste and consistent definitions of different types of materials. Above all, legal frameworks need to be comprehensive and coherent, addressing the product life cycle in its entirety. In order to effectively promote the creation of business opportunities from sustainable waste management practices, legal and policy frameworks must also encompass sectors other than environmental protection, including for example commerce and taxation. Possible tools include the use of economic instruments such as tax incentives and disincentives; promotion and use of third-party environmental health and safety certification standards as a means of identifying environmentally sound operators and facilities; and minimization of barriers to trade within countries as well as internationally. This will require rethinking of the relationships between environmental and trade legislation both nationally and internationally. An approach that deserves to be further explored is adopting international standards and certification schemes for specific facilities and processes, and permitting resource and energy recovery from particular waste materials (including transboundary movements for this purpose) only in certified facilities and through certified processes.[12]

A starting point could be an overarching international treaty that would encompass the entire life cycle of materials management and remedy the gaps that have been identified in this book. Such a framework treaty could embody unified basic principles of materials management in a 'cradle-to-cradle' approach, and thus provide a frame of reference within which national and regional differences could be taken into account. The existing chemicals and waste treaties, including the Basel Convention, could operate within this framework, and protocols on additional materials could subsequently be adopted as necessary. Alternatively, the framework treaty could oblige its parties to elaborate national legislation on particular

---

[12]   Ibid., 347.

substances and aspects of their management, based on the fundamental principles of the treaty. To facilitate this, and to ensure maximum consistency among the resulting national laws, a set of guidelines could be developed under the framework treaty.[13]

Such unification could be an important and innovative first step in creating a secure investment environment. However, creating a new international legal framework that adequately addresses the many cross-cutting issues (trade, environment, health, human rights etc.) remains a very difficult endeavour in treaty drafting, even assuming the existence of a corresponding political commitment of the international community.

Finally, it must be borne in mind that the existence of a solid regulatory framework is only a first step towards an environment that would make creating business opportunities from environmentally sound waste management while protecting human health and the environment a reality: this will also require consistent implementation and enforcement of the laws that are in place. This in turn depends on the existence of an effective overall governance framework at the national and sub-national levels, based on the rule of law.

---

[13] For a more in-depth discussion of this approach see Kummer Peiry (n 1) 178.

Index